INTRODUCTION

Floor of Meteor Crater in November 1972, with Mr C F Lewis of Arizona State University (right) and Dr F Burragato of the University of Rome. Early this century an attempt was made to locate and exploit the iron-nickel asteroid thought to be buried beneath the crater. In the 1920's it was realised that the projectile must have vaporised on impact, so the drilling equipment was abandoned.

Meteorites are pieces of asteroids, pieces of the Moon, of Mars and perhaps even of comets, that fall to Earth from space. Their arrival may be heralded by a ball of light followed by sonic booms. More rarely, their impact may be explosive enough to alter the course of Earth history.

Some meteorites are similar to volcanic rocks on Earth. Some are alloys of iron and nickel similar to the material which is is believed to form the Earth's core. Many others, however, are like nothing on Earth and may represent the primitive material from which our Solar System was made.

Study of these primitive meteorites, in the laboratory, enables scientists to handle some of the stardust from which the Solar System – the Sun, the Earth, and even ourselves – was formed. This is something of a paradox, because meteorites are often, wrongly, called 'falling stars'.

Meteorites are samples from parts of the Solar System astronauts may never be able to visit, or which would cost a great deal to explore. They have been dubbed 'the poor man's space probe'. Cheapness of meteorite recovery contrasts sharply with the wealth of information which they provide. Meteorites have enabled us to come close to answering questions such as 'How did the Earth, Sun and planets form?', and even, 'How did we come to

WHAT ARE METEORITES?

Meteorites are natural objects that survive their fall to Earth from space. A meteorite is usually named after a place near where it was seen to fall, or was found. Meteorite literature is full of exotic names such as Campo del Cielo (Argentina), Prairie Dog Creek (USA) and Zagora (Morocco).

Most meteorites are lumps of stone, usually with a small amount of iron-nickel metal. Others are almost entirely composed of iron-nickel metal, so are very different from familiar Earth rocks.

We normally think of a meteorite as a single object weighing anything from a few grams to a few tonnes, but in some cases a meteorite fall comprises thousands of fragments of stone or metal totalling many tonnes. Such multiple falls are known as meteorite showers. Each fragment in a meteorite

1 The Middlesborough meteorite that fell on March 14, 1884, in Yorkshire, England. This stony meteorite has a conical shape with a diameter of 15 centimetres (6 inches). The apex (centre of figure) of the cone pointed towards the direction of travel during atmospheric flight. The flutings that radiate from the apex were produced by atmospheric ablation.

2 An iron meteorite, weighing 634 kilograms, from the Campo del Cielo, Argentina, shower. Many masses of meteoritic iron have been found in the area since 1576, and some tens of tonnes have been recovered.

1

2

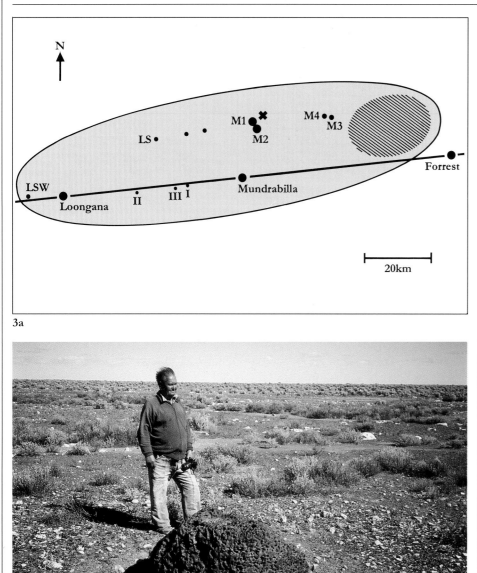

3a

3a Map of the strewn field of the Mundrabilla, Western Australia, iron meteorite shower. The straight line through Mundrabilla and Loongana represents the Trans-Australian Railway. The meteoroid broke up during atmospheric flight. Two masses of iron of 11 and 6 tonnes (M1,M2) were found by a survey party in 1966, and other smaller meteorites were attributed to the same shower. In the shaded area many small masses, each weighing up to a few hundreds of grams, were found.

3b A 3.5 tonne mass of the Mundrabilla shower, with John Carlisle, the finder. He has spent much of the last fifty years roaming the Nullarbor Plain and has found many meteorites. This is his largest single find (at 'x', Fig. 3a). The photo was taken by R Hutchison in August 1988, just before the mass was taken to the Western Australian Museum, in Perth.

shower is known as an **individual**.
When a large incoming object does not break up in the atmosphere but hits the Earth intact, it produces an explosion crater on impact.

3b

3

MICROMETEORITES

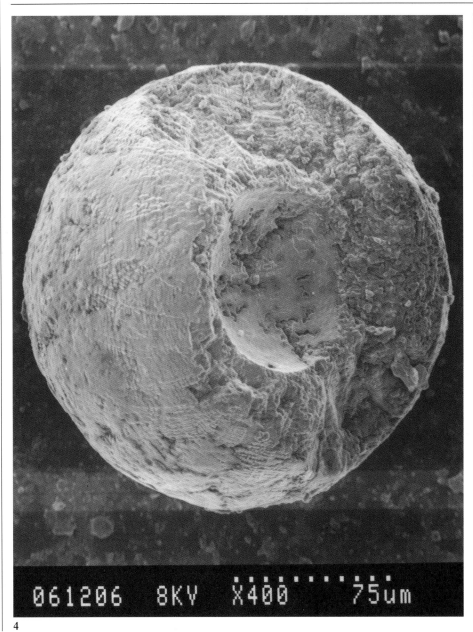

061206   8KV   X400   75um

The smallest objects coming to Earth from space are cosmic spherules and interplanetary dust particles. Cosmic spherules are tiny droplets from space about one millimetre or less in diameter. They have been found in the clay beneath the oceans remote from land and in ice from Antarctica and Greenland.

Interplanetary dust particles are irregular in shape and less than one-twentieth of a millimetre in size. From the late 1970s they have been collected by high flying aircraft operated by NASA. Like cosmic spherules, these particles too have recently been discovered in Antarctic ice.

Cosmic spherules and interplanetary dust particles are together called **micrometeorites**. Their small size allows heat to radiate away from them, so they are not melted during their passage through the atmosphere. They retain the form they had in space. Large meteorites are mineralogically different from micrometeorites, many of which probably come from comets.

4 A cosmic spherule, 0.2 millimetres in diameter, found in clay dredged from the bottom of the Pacific Ocean. The hemispherical pit in the middle of the picture is thought to have been caused by the loss of a globule of iron-nickel metal during the solidification of the spherule.

4

5a Interplanetary dust particle, 20 micrometres (one fiftieth of a millimetre) across. A friable aggregate of many mineral grains that may have come from a comet.

5b Aircraft operated by NASA (U.S. National Aeronautics and Space Administration) to collect dust in the stratosphere. When the aircraft climbs to about 20 kilometres altitude (about 60000 feet), panels covered in silicone grease are extended. The plane cruises for a time and dust sticks to the grease. On descending, the panels are retracted to avoid contamination by terrestrial dust. The panels are taken to a clean laboratory where the silicone grease is dissolved away to leave the interplanetary dust trapped on filters.

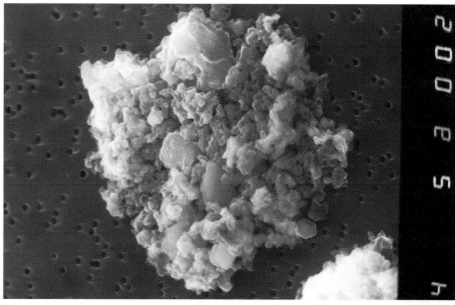

5a

5b

WHERE DO METEORITES FALL?

6

The fall of meteorites on Earth is largely random. Meteorites are found from the tropics to the poles.

If a meteorite lands more than a few hundred metres from a witness, the chances are that it will not be found: so in uninhabited areas such as polar regions no meteorite fall has yet been observed. Meteorites that hit vehicles or buildings are obvious exceptions. Even when its fall was not witnessed, the meteorite may be found later because of the damage to property.

There is no reliable record of anyone having been killed by a meteorite, although the Nakhla (Egypt) fall of 1911 killed a dog. A number of people were slightly injured by the blast from the Tunguska (Siberia) explosion of 1908, and a woman was struck by the Sylacauga (Alabama) meteorite in 1954.

Theory predicts that for every three meteorites that land near the poles, four land near the equator. This is because gravitational focussing favours equatorial falls (Fig. 8).

Meteorites can fall at any time of day, but most fall in the afternoon or evening. Such falls are favoured because, as the Earth spins, the hemisphere with noon to midnight local time encounters objects in orbit around the Sun that are closing in on the Earth. In this case, capture by the Earth's gravity is more likely. On the contrary, objects that pass in front of the hemisphere with local time

7

6 Hoba iron meteorite. This, the biggest single meteorite known, is still in the ground in Namibia. Its estimated weight is 60 tonnes, but part had rusted away so that its original weight may have been 100 tonnes. The photo was taken in the 1920s; second from left is Dr L.J. Spencer, who became Keeper (Head of Department) of Minerals at the British Museum (Natural History).

7 Dr Bill Cassidy discovering a small stony meteorite (a rare achondrite) on blue ice in Antarctica. Even tiny meteorites can be spotted from a helicopter flying over the ice. Most, however, have been found by scientists on the ground

between midnight and noon are pulling away from the Earth. In this case gravity is less likely to pull in an object (Fig. 9).

Most bodies travel around the Sun in the same sense as the Earth, so the difference in speed between the Earth and such bodies during an encounter is small, that is, only a few kilometres per second. The slower the speed on encounter, the greater the chance of material surviving to the Earth's surface rather than burning up in the atmosphere. The rare bodies travelling around the Sun in the oposite direction to the Earth tend to hit, head on, the hemisphere with time between midnight and noon. These encounters may occur at 70 kilometres per second, so material has little chance of survival.

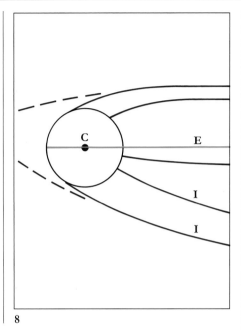

8

8 Sketch of the paths of meteoroids relative to the Earth. CE is the *ecliptic*, the plane in which the Earth (with centre C) orbits the Sun. Meteoroids (I) inclined to the ecliptic are deflected, along the solid lines, by the Earth's gravity (but less so if the velocity is high – dashed lines). This *gravitational focussing* generally concentrates falls towards the equatorial region of the Earth. Very occasionally it can result in falls on the side of the Earth facing away from the direction of travel of the meteoroid. Because of gravitational focussing, the Earth captures matter from a zone that is wider than its diameter.

9 Meteorite falls and time of day. Most meteorites (M) are pieces broken from asteroids, and travel around the Sun in the same sense as the Earth. At the same distance from the Sun as the Earth, pieces of asteroid travel faster in their orbits, otherwise they would not move further from the Sun. As the Earth spins, the half that faces backwards along its path has local time from noon to midnight. Pieces of asteroid coming from this direction catch up with the Earth, as shown in the figure, so are more easily captured than those that pass in front of the Earth, which pull away. Thus, there are more meteorite falls on the backward-facing hemisphere, with local time from noon to midnight, than on the half with time from midnight to noon. Rare objects in **retrograde** orbit (not shown) collide head-on with the half with time from midnight to noon. Such material seldom, if ever, survives.

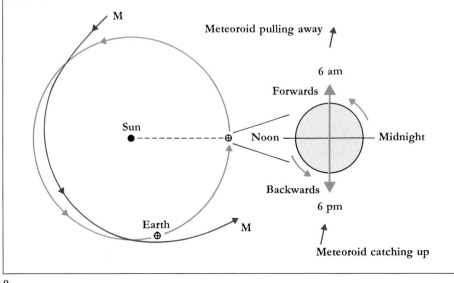

9

HOW DO METEORITES GET HERE?

Bodies in orbit around the Sun are called **meteoroids** if there is a chance that they may at some time land on Earth.

On encountering the Earth, a meteoroid's velocity is always more than 11.2 kilometres per second (the Earth's **escape velocity**). This is about forty times faster than the speed of sound. When the meteoroid is at a height of 100-120 kilometres, frictional heating in the upper atmosphere can produce a bright fireball.

What happens next depends on the speed, mass and friability of the meteoroid. Very small objects burn up as **meteors** or **shooting stars**, but the tiniest may survive to become micrometeorites.

Friable meteoroids break up and produce multiple trails as they are destroyed when they are still 80 or 90 kilometres up.

Tough objects weighing more than about 10 grams produce fireballs and some material may survive to lower altitudes. Larger objects produce fireballs that may rival the Sun in brightness. During its passage through the atmosphere, a meteoroid less than one metre in diameter may be enveloped in a moving ball of incandescent gas of some 200 metres diameter. This is caused by melting and vaporization of the surface of the meteoroid by frictional heat and by the passage of the object giving the air an electric charge. *All* the heat is carried away from the meteoroid by the melt

and vapour, while the interior stays cold. As the meteoroid penetrates to lower altitudes, the denser air causes increased deceleration. If material survives to subsonic speeds, the fireball is extinguished and the residue falls under gravity to become a meteorite. The last melted material on the surface of the body solidifies to form a thin, usually black, **fusion crust** (see Fig. 14).

A bright fireball may be visible for hundreds of kilometres, often appearing to have a range of colours. It flares and dims as the meteoroid disintegrates and appears as a fast moving, glowing ball with a tail, except to observers directly in its path. In this case the tail is invisible and observers see an apparently stationary fireball that gets bigger as it moves towards

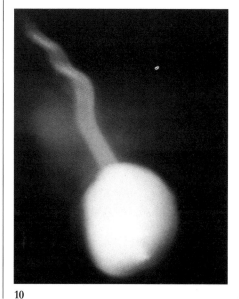

10

11

them. Fireballs are normally visible for only a few seconds, but a dust trail (or 'train') may be left in the sky, for tens of minutes.

Sonic booms often follow the appearance of a fireball, just as thunder follows lightning. In addition, mysterious sounds of crackling or hissing may accompany a sighting simultaneously. The phenomena causing the sounds must travel with the speed of light, so they are thought to be linked to electromagnetic radiation (such as light). The sonic effects from bright fireballs can cause windows to rattle or break, and the ground may shake. Afterwards, humming or whistling sounds may signal the fall of objects which then become meteorites.

12

10 Fireball of the Pasamonte meteorite at 5.00 a.m. on 24 March 1933. Note the twisting in the dust trail indicating that the meteoroid spiralled in its flight.

11 Meteors in the atmosphere. The Leonid shower of November 1799, from near the coast of north-east USA. At set times each year the Earth passes through debris distributed unevenly around the orbits of comets.. The uneven distribution causes meteor showers to vary in intensity from year to year In 1799 the Earth passed through a particularly dense part of the Leonid meteor stream, to produce this awesome display. The meteors are known as *Leonids* because they appear to radiate from the constellation of Leo.

12 Meteor of 18 August 1783, seen from near Newark upon Trent, England. A single fireball that quickly broke up into many small ones was seen.

13 A summary of atmospheric entry of different types of object. Tiny or friable particles burn up high in the atmosphere as meteors. Larger, friable or high-velocity objects survive to lower altitudes, but also are destroyed. Tough, low-velocity objects are decelerated and fall under gravity (dashed lines) to become single meteorites or meteorite showers. Break-up usually occurs 10-30 kilometres up, between the dashed horizontal lines. Large, tough objects are not significantly decelerated and produce explosion craters.

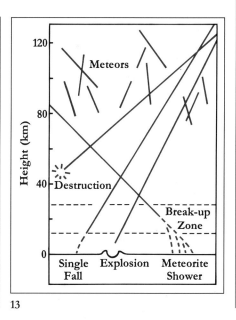

13

HOW DO METEORITES GET HERE?

Atmospheric phenomena differ greatly from one fireball to another. For example:

THE ALDSWORTH, GLOUCESTERSHIRE, ENGLAND, FALL OF 4 AUGUST 1835

An account, published in 1852, states that at about 4.30 p.m. a loud boom followed by a low rumbling was heard over several English counties in the Welsh borderlands. A fireball that looked like a copper ball larger than an orange was seen from Cirencester. The fireball passed from west to east, had a tail and made a rumbling noise like thunder that was heard by many people. The inhabitants of Cirencester marvelled at the thunder on a serene day with a cloudless sky. At Aldsworth some 20 kilometres (13 miles) to the north-east, a meteoritic stone fell within 20 metres of workmen in a field. They saw *'no unusual light, but heard the aerolite* (stony meteorite) *rush through the air and felt it shake the ground by striking it with great violence.'* It drove *'straw before it down into the earth for six inches* (15 centimetres)*, till opposed by rock. When the men got it up it was not hot.'* About 800 metres south of where the stone fell there was *'a shower of small pieces . . . Children thought it was a shower of black beetles, and held out their hands to catch them as they fell.'* This fall is unusual in that part of the dust trail landed soon after the stone which weighed 600 grams. It is now preserved in The Natural History Museum, London.

14

14 The Aldsworth meteorite (about 10 centimetres across). The stone was cut (top left) to obtain specimens for research, but much of the remainder is coated with dark grey *fusion crust*. When a meteoroid enters the atmosphere at great speed, friction causes the surface to melt. Molten droplets are carried into the atmosphere and the surface of the body is eaten away, or *ablated*. Friction slows the meteoroid and melting of the surface stops. Melt still on the surface freezes to a glassy crust, as on Aldsworth.

15

THE PASAMONTE, NEW MEXICO, USA, FALL OF 24 MARCH 1933

The spectacular fireball illustrated in Fig. 10 was seen by observers over parts of New Mexico, Colorado, Kansas, Oklahoma and Texas. It appeared at 5 a.m. on 24 March 1933 and had a low-angled trajectory somewhat south of west. The event was thoroughly investigated by Dr H H Nininger who devoted over 60 years of his life to the recovery and study of meteorites. Nininger, who died on 1 March 1986, aged 99, personally interviewed witnesses over an area of about 300 000 square kilometres (300x400 miles), so was able to deduce where the meteorite landed. A search over the following year led to the recovery of a shower of meteoritic stones.

The light was so bright that it startled a ranch foreman who was indoors and about to sit down to breakfast. Luckily, he was able to pick up a loaded camera and go outside to take the photograph.

Some observers saw only one fireball, others saw a cluster, or procession, of fireballs and a few people witnessed the break-up of a single fireball. To some, the fireball appeared to change direction, which is consistent with the twisting dust-trail in Fig. 10. Many observers heard 'a swishing or whining noise at the instant of the fireball's passage' – an example of the 'mysterious sounds' already described. Sonic booms were heard two or three minutes later. A most unusual feature was the presence of a luminous cloud, some hundreds of kilometres long. The luminosity persisted for up to half an hour after the fireball, but the cloud could not have been lit by the Sun because it was too early in the morning.

15 The luminous dust cloud after the passage and fall of the Pasamonte meteorite, 24 March 1933.

HOW DO METEORITES GET HERE?

16a

THE JILIN, CHINA, FALL OF 8 MARCH 1976

This fall produced the largest known stony meteorite. After a bright fireball was observed at about 3 p.m., a shower of stones totalling four tonnes fell. Fortunately, no one was injured. The largest stone weighed 1.77 tonnes and made an impact pit six metres deep.

16a & 16b The impact pit of the largest known individual stony meteorite (1.77 tonnes) before and after excavation. The stone is part of a meteorite shower that fell at Jilin, China, on 8 March 1976.

16b

WESTERN AUSTRALIA, 10 SEPTEMBER 1990, 2305 WESTERN AUSTRALIA STANDARD TIME

The following report illustrates the information now deemed important in fireball observation. No meteorite has yet been recovered, but it is highly likely that some material survived to land on Earth. The text is from the *Bulletin of the Global Volcanism Network*, which also covers earthquakes, fireballs and meteorite falls. It is published by the Smithsonian Institution, Washington, DC, USA.

Location: Central W coast and adjacent agricultural region of W Australia.

Observers: A total of 343 reports were received by the W Australia observatory, Bureau of Meteorology, police and media. Observations of the fireball were collected from a large area of Western Australia, extending N from Perth and including Denham, Koorda and Cue.

Magnitude: The fireball was very bright with a **magnitude** estimated . . . . at greater than -16. It was too bright to look at directly and "turned night into day".

Color: Violet, blue, and white. Red sparks and a red halo were observed.

Train: Lasted about 90 seconds after the fireball had disappeared. The train was noticeably contorted by upper atmosphere winds, and drifted SE.

Trajectory: Detailed trajectory information is not available, but it is believed that the meteorite may have fallen near Yuna (28°20'N, 115°00'E) about 400 km N of Perth.

Sonic Effects: Electrophonic sounds including whistles and crackles were heard while the fireball was visible; sonic booms were heard 60 – 90 seconds after it disappeared. All the reports of sonic effects were from the Geraldton area, 360 km N of Perth.

Information Contact: Mr or Ms X, Balajura, Western Australia.

THE GLATTON, CAMBRIDGESHIRE, ENGLAND, FALL OF 5 MAY 1991

This was the first observed British fall for over twenty-two years. As Mr Arthur Pettifor worked in his garden on a cold, overcast day, he was startled by a loud whining noise, followed by the crash of an object through his conifer hedge some twenty metres away. Mr Pettifor found a single stone of 767 grams that was warm, not hot, to the touch. No fireball was seen and no sonic booms were reported. Had Mr Pettifor been indoors, the fall probably would not have been noticed.

One week after the fall neighbouring farmland was searched but no additional stones were found. Measurements indicated that the object, in space, had been no bigger than a large grapefruit.

17a Mr Arthur Pettifor holding the Glatton meteorite.

17b The Glatton meteorite was almost completely covered with very dark fusion crust, but the light grey interior is revealed at the upper corner where the crust had broken off. The meteorite proved to be stony and of the type of chondrite most commonly seen to fall.

17a

17b

WHERE DO METEORITES COME FROM?

We know from their association with bright fireballs that meteorites come from space, but to identify the source regions we need to determine the orbits of the meteoroids when they encountered the Earth. This requires knowledge of the direction and speed of each meteoroid during atmospheric flight. The rough direction can be deduced from at least ten sightings by reliable observers distributed on both sides of the flight-path. Speed, however, can only be measured from photographs taken by specially designed cameras. We have this type of information for three meteorite falls only.

The bright fireball from the first of the three was photographed by chance by cameras set up for meteor (shooting star) photography. The Pribram fireball of 1959 was so bright that the plates were almost burned out, but they were salvaged and used successfully in predicting the area to search for a meteorite. Camera networks were subsequently set up in the prairie states of the USA and in Canada, and each has had a single success.

From the photographs the orbit of each meteoroid was worked out. All three orbits were found to be elliptical and to extend from the Earth's orbit to the asteroid belt, between Mars and Jupiter. Such orbits are typical of those of a family of **asteroids** (Apollo asteroids).

The asteroid belt contains innumerable tiny planets or planetary fragments in stable, near circular orbits

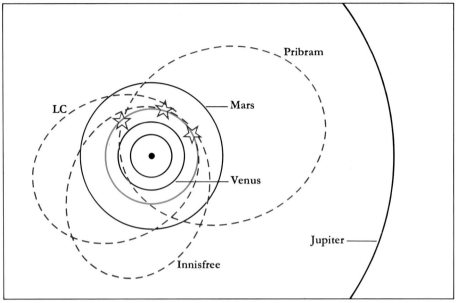

18

around the Sun. Sometimes Jupiter's gravity pulls an asteroid into an elliptical orbit. If the elliptical orbit crosses the Earth's orbit, the asteroid (or fragments from it) may land on Earth.

Some objects, in orbits like those of Apollo asteroids, appear to be fuzzy when viewed by astronomical telescopes. These objects are comets with short periods (comets which orbit in the inner Solar System only and so are seen more frequently), and it is clear that they, or their debris, also land on Earth. Long-period comets have highly elliptical orbits and their velocity on encountering the Earth is so great that frictional effects in the atmosphere would result in their total, and perhaps explosive, destruction. It may, however,

18 Orbits of the Pribram (Czechoslovakia), Lost City (USA) and Innisfree (Canada) meteorites, the only three that have been determined precisely from photographs. The orbits are elliptical and extend beyond the orbit of Mars, but not as far as Jupiter's orbit. The meteorites landed on Earth so their paths cross the Earth's orbit, but only one extends inwards as far as the orbit of Venus.

be possible for a long-period comet to be captured into an orbit in the inner Solar System and so become a short-period comet. So, we probably receive both cometary and asteroidal material as both meteorites and micrometeorites.

In addition to asteroidal and

WHERE DO METEORITES COME FROM?

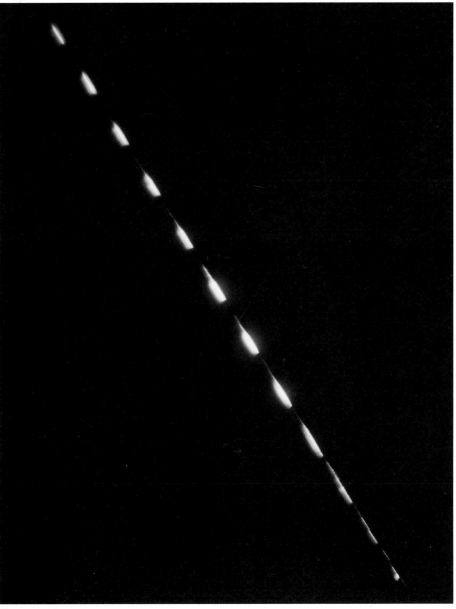

cometary sources, the compositions of twelve meteorites indicate that they are from the Moon and eight are probably from Mars.

No observed fireball, meteor or comet has had a speed greater than the escape velocity of the Solar System. For this and other reasons we believe that all meteorites originate within the Solar System. The escape velocity of the Solar System is the speed that an object must have, on leaving the Sun, for the object to overcome the Sun's gravity and escape from its gravitational field to enter interstellar space. Conversely, an object moving from interstellar space towards the Sun will be pulled by it when the Sun's gravitational field is entered. The speed of the object will be its speed in interstellar space **plus** the speed caused by the Sun's gravity – the escape velocity. Thus, anything coming into the Solar System from outside will travel faster than the escape velocity. Because this has not been observed we know that meteorites landing on Earth belong to the Solar System, but some meteors come from outside.

19 Path of the fireball of the Innisfree, Alberta, meteorite of 5 February 1977. The picture was taken by an automatic camera of the Canadian Meteorite Observation and Recovery Programme. A rotating wheel produces four segments each second. The meteoroid fragmented towards the end of its path (lower right), and the fragments then travelled at different speeds

HOW MANY METEORITES LAND ON EARTH?

Fireballs observed by the Canadian Camera Network suggest that 26 000 meteorites, each weighing over 100 grams, land on Earth each year. Most fall in the oceans which cover 70 per cent of the Earth's surface. Of the remainder only five or six falls are witnessed or damage property, so are recovered and become known to science. Specimens of some 900 observed **falls** are represented in collections, and most fell during the past two centuries.

Apart from meteorites, meteors and micrometeorites contribute perhaps 10 000 tonnes of material to the Earth annually. This is not a significant addition to the mass of the Earth (5.98×10^{21} tonnes) over the 4550 million years of its existence.

Many meteoroids break up in the atmosphere and produce multiple falls. As a meteoroid penetrates into the dense, lower atmosphere at hypersonic speed, the air is compressed at the front while a vacuum forms behind. The object is stressed and fracture often occurs, especially if there are weaknesses within it. Such weaknesses may be cracks or shock-veins caused by collisions between asteroids in space.

Fragmentation may take place in a series of stages to produce a meteorite shower comprising a few large pieces and many small ones. Most of the weight, however, is usually concentrated in the large lumps. The individuals of each shower land within an elliptical area on the Earth's surface, with the largest masses at the end furthest along the line of travel. (See Fig. 3a.)

20 Woodcut depicting the fall of the Ensisheim meteorite in Alsace, France, on 16 November 1492. The suggestion that the meteorite was associated with black clouds, thunder and lightning is in fact erroneous – a touch of artistic licence.

21 One stone from the Cronstad (South Africa) meteorite shower of 19 November 1877. Note the abundant, shiny metal grains on the saw-cut surface, which is about 8 centimetres across. The meteorite is *brecciated*, that is, composed of angular fragments, in this case welded together to form a tough rock. The junction between one large fragment and several darker ones can be seen at the left of the cut face.

20

21

Most of the meteorites in collections are **finds** – objects that have been identified as extraterrestrial but whose arrival on Earth was not witnessed. Until recently, most were found in deserts where the low rainfall causes only slow weathering. For example, over 100 meteorites have been found in one area of New Mexico, and more have been discovered on the Nullarbor Plain, in Australia.

In contrast, only one meteorite has been **found** in Britain, the other 20 being observed falls. This find came from an Iron Age site at Danebury Hill, in Hampshire, that was occupied from 800 BC to 50 BC. Many pits, some 1.5 metres in both depth and width, were dug by the former inhabitants and the meteorite was found in 1974 during the excavation of one pit, but was not studied scientifically until 1989. It weighed only about 30 grams and although partly rusted, proved to be of a common stony type. The presence of meteorites in archaeological sites elsewhere, such as the burial mounds of the Hopewell culture in the USA, indicates that these objects were commonly revered by primitive tribes. We cannot say, however, whether the Danebury meteorite was placed in the pit or merely fell there.

The almost complete lack of 'finds' in Britain may be ascribed to the moist climate, long history of farming, and intense glaciation that deposited boulders over much of the surface. It

22

is unlikely that a stone from the sky would have been recognized as unusual if it were associated with glacial debris from various sources.

Most iron meteorites are finds rather than observed falls. (Only 4 per cent of observed falls are of iron meteorites.) They resist weathering, are very distinct from Earth rocks, and tend to be picked out as unusual.

22 Dr Alex Bevan (Curator of Minerals and Meteorites, Western Australian Museum) at a camp near Camel Donga on the Nullarbor Plain. The bag is full of specimens of the Camel Donga meteorite shower. The expanse of bare gravel and sparse vegetation indicates a low rainfall, so that weathering of meteorites is minimal.

ANTARCTIC METEORITES

In 1969 nine meteorites were found by chance by a Japanese party working in Dronning Maud Land, Antarctica. When the meteorites were studied they proved to be of different types, which was wholly unexpected. A group of meteorites found together would normally be from a single fall, so they would have identical compositions. It is highly improbable that nine meteorites of such different types were associated together in space, so these meteorites must have come from different sources and have fallen at different times. Consequently, in Antarctica there must exist special conditions which preserve and concentrate meteorites from different falls. During late 1973 and in every brief Antarctic summer (December and January) thereafter, expeditions have gone to Antarctica to recover meteorites.

The meteorites are generally found in 'blue ice' areas. Here snow from the interior ice cap, after compression into ice, but still with its cargo of material from space, rises around mountain barriers near the coast. Fierce winds strip the ice from the surface, exposing the blue ice from depth, together with its meteorites which become

23 Cross-section of blue ice area, Antarctica. Ice builds up on the high polar plateau, and the pressure forces it to move northwards, towards the coast. In some cases, as shown, the ice encounters a mountain barrier, which slows its movement. Fierce winds coming down from the pole blow the ice away, and cut down to lower layers, forming a depression. Here, the surface of the ice is an erosion surface. As the load of overlying ice is eroded by wind, older ice layers rise to take its place. This deep blue ice carries its meteorites to the surface, where they accumulate.

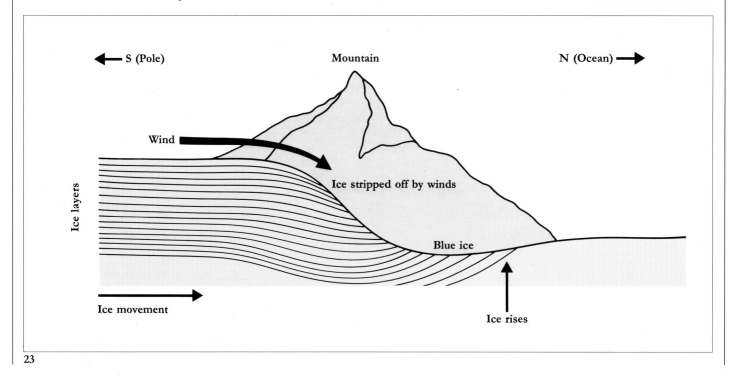

S (Pole)   Mountain   N (Ocean)

Wind

Ice stripped off by winds

Ice layers

Blue ice

Ice movement

Ice rises

concentrated on the erosion surface.

From satellite images, 'blue ice' areas have been identified in other parts of Antarctica. Since 1976, US expeditions have successfully searched for meteorites in Victoria Land and a European party operating in a different area has just found 264 meteorites (January 1991).

Over 10 000 individual specimens of meteorites have been recovered from the continent, but many of these must be from the same meteorite showers. Because of the presence of rare and unusual types, it is likely that more than a thousand falls are represented and the number of meteorites known to science has increased from about 2000 to 3500 in only 20 years.

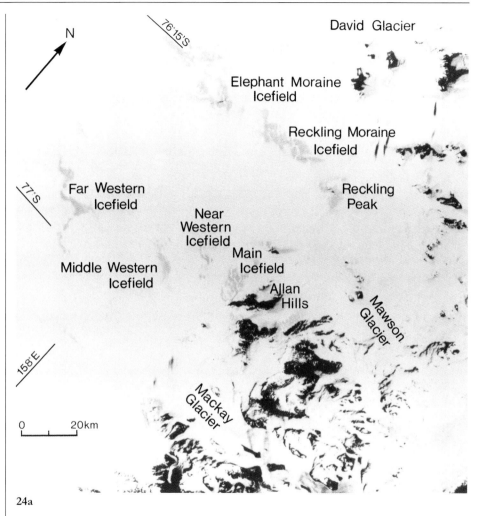

24a

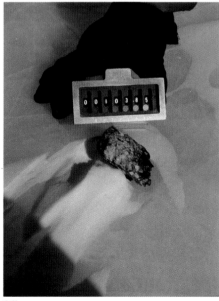

24b

24a Satellite image of part of Victoria Land, Antarctica. Snow cover appears as white, rock outcrops black. Pale grey areas are of bare, blue ice on which many meteorites have been found, especially to the west of the Allan Hills.

24b An achondritic stony meteorite on blue ice of the Allan Hills area, Victoria Land, Antartica. The counter displays the field identification number and a scale in centimetres. The stone was taken to NASA's Johnson Space Center for study.

WHAT ARE METEORITES MADE OF?

Meteorites are composed of **minerals**, most of which also occur in rocks on Earth. Minerals are the 'building blocks' of rocks and are ordered arrangements of chemical elements, such as iron, oxygen, aluminium and sodium. All known elements are present in meteorites, although some, such as hydrogen and helium, are very rare compared with their great abundances in stars, or in the Universe as a whole. Meteorites, then, are mainly composed of commonplace materials, but in proportions that differ from those in the accessible rocks on Earth (Fig. 25). Rocks similar to some meteorites may well occur within the interior of the Earth.

On the basis of their composition, meteorites are divided into **stony**, **iron** and **stony-iron** types. Stony meteorites, which, as their name suggests, are dominantly made of stony minerals, are further subdivided on the basis of their texture into **chondrites**, which contain small, near spherical **chondrules**, and **achondrites**, which are similar to some rocks found on Earth.

Almost all chondrites contain some iron-nickel metal in addition to stony minerals. From their textures and because of the mixture of stone and metal, which would separate on melting, we know that chondrite meteorites have not melted after their formation. In contrast, metal is rare or absent in achondrites. We think that melting on a small planet or asteroid composed of a mixture of stony and iron minerals caused metal to sink and molten stony liquid to rise, like iron and slag in a blast furnace. The metal became concentrated in asteroidal (or planetary) cores, the source of many iron meteorites, and the stony liquids cooled to become **igneous rocks** similar to basalt lava on Earth. The achondrites are formed of these igneous rocks. Some igneous rocks near the surfaces of their parent asteroids were broken up by impact, to form 'soils'. These 'soils' also occur among the achondrites.

25 Cross-section through a differentiated planet, like the Earth. Soon after its formation, the release of heat caused a planet to melt. The primordial mixture of stony and metallic minerals formed melts of different density. Dense, iron-nickel metal and iron sulphide liquids sank towards the centre to form an inner and outer core, respectively. Less dense, stony liquids were left behind to form a mantle around the core. Unmelted rubble on the surface, together with later basaltic melts from the mantle, formed the crust.

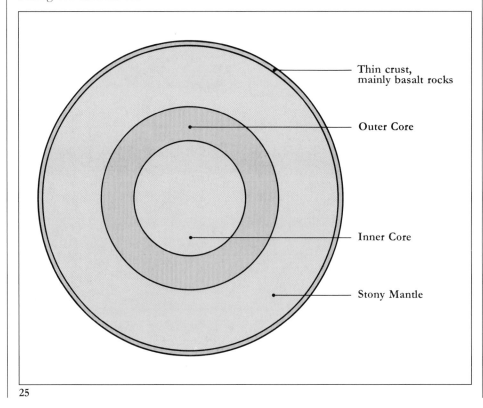

Thin crust, mainly basalt rocks

Outer Core

Inner Core

Stony Mantle

25

STONY METEORITES: CHONDRITES

Almost 90 per cent of the meteorites seen to fall are chondrites, as are most of those found in Antarctica. These are stony types that, with few exceptions, contain some metal and sulphide. Amost all contain chondrules – near spherical, millimetre-sized, mainly silicate (stony) objects, many of which formed by rapid cooling and solidification of molten droplets. To melt such materials requires a temperature of about 1400°C and this shows that there must have been at least one large scale heating event in the early Solar System.

Although the chondrules within the chondrite meteorites are solidified melts or abraded fragments of igneous rock, the melting occurred before the various components aggregated to form the chondrites. Chondrites have not been melted after their formation.

Because they have not been melted since their formation and because they are a mixture of stony, metallic and sulphide minerals, chondrites are believed to represent the materials from which the Earth formed some 4550 million years ago. In fact, the chemical composition of chondrites closely matches that of the Sun, if the gaseous elements such as hydrogen and helium are ignored.

There is however, much chemical variation among the chondrites. A few contain water and 'organic' compounds, but almost no iron-nickel metal.

26a

26a Cut and polished face (7.5 centimetres across) of the Beddgelert chondrite, that fell in Wales in 1949. Shiny grains of iron-nickel metal and larger, grey to cream coloured, sub-circular chondrules are evenly distributed.

26b Cross-section through a spherical droplet chondrule (2 millimetres in diameter), photographed with polarized light. It was originally molten at about 1500°C. From the rapidly cooled melt, several imperfect, branching crystals of olivine (grey to orange) grew, but the remainder of the liquid cooled to a glass (black). The glass and the structure of the crystals indicate rapid cooling.

26b

WHAT ARE METEORITES MADE OF?

In contrast, others contain over 20 per cent of metal and tiny amounts of rare oxygen-poor minerals that do not occur naturally on Earth. One of these is niningerite, a magnesium-iron sulphide, named in honour of Dr H H Nininger. Most chondrites lie between these extremes and contain between 2 and 20 per cent iron-nickel metal but little water or carbon (see Fig. 26a).

Many scientists believe that the components of chondrites were produced in a gas-dust cloud or 'pre-solar nebula' from which the Solar System formed. One popular theory is that condensation in the nebula gave rise to dust-balls which were melted by collision, lightning or some other means and became chondrules. The chondrules cooled and stuck together, with trapped dust, and ultimately grew into planets. Chondrite meteorites are small fragments of these small planets.

Other scientists argue that the chondrites do not record the earliest planet formation. Some chondrites contain rare rock fragments that have undergone melting on a planetary scale. Planets, it is argued, must therefore have existed before the chondrites formed. This suggests that chondrules and chondrites are the product of collisions between planets which may have been only tens or hundreds of kilometres in diameter. These same chondrites contain tiny amounts of dust that formed in the atmospheres of stars, so the meteorites do retain a record of a nebular phase.

27 Chondrules of the Sharps, USA, chondrite, under the microscope with polarized light (2.5 millimetres across). The grey and white, round chondrule is composed mainly of interlocking, elongate imperfect crystals of the mineral, pyroxene. The elongate object is composed of two feathery, interlocking crystals of olivine, one being mauve, the other, yellow. These textures indicate that the chondrules formed by rapid cooling of molten droplets. The round chondrule is surrounded by a dark rim that formed by the adhesion of dust.

28 Cut and polished surface (6 centimetres across) of the Khairpur meteorite that fell in 1873 in north-west India (now Pakistan). Shiny metal is abundant in this oxygen-poor chondrite.

27

28

STONY METEORITES: CARBONACEOUS CHONDRITES

Only about ninety meteorites are included in this general category, but although small in number they provide a great amount of varied information on the origin of the Sun and planets, and of life itself. Most carbonaceous chondrites contain chondrules, and other objects formed at high temperature, set in a matrix with clays that contain water (fig 29).

Within the matrix are 'organic' compounds of carbon, hydrogen, oxygen and nitrogen, the elements that are the main constituents of living cells. Some of these compounds are exceedingly rare in living organisms on Earth, so it is thought that the compounds in the meteorites were formed in space and that life was not involved. The organic compounds in meteorites are important as an indication of what must have landed on Earth before life began. From them we can determine possible pathways from which life might have evolved.

29 A stone of the Orgueil, France, meteorite shower that fell in 1864 (10 centimetres across). It is one of five or six meteorites with about 20 per cent water (by weight) and no metal. The dark crust is flaking off to reveal water-bearing clay-like minerals and whitish carbonate and sulphate. Some carbon occurs as compounds with hydrogen, oxygen and nitrogen that are of interest in studies of the origin of life.

29

WHAT ARE METEORITES MADE OF?

The matrix also contains tiny crystals of diamond and silicon carbide (carborundum). These minerals have been implanted with gases, which indicates that they formed in the atmospheres of massive stars, called **red giants.** Matter ejected from the red giants contributed to the gas-dust cloud from which the Sun and planets formed. It is an exciting thought that scientists can isolate star-dust in the laboratory and so study processes that occurred tens or hundreds of millions of years before the origin of the Solar System and our Earth.

Carbonaceous chondrites contain rare calcium and aluminium oxide minerals that can survive great heat. These minerals occur in clumps that may have been the first material to condense from a hot nebula just before the planets formed. They provide a window into the earliest events in the origin of the Solar System, after the diamond had been formed.

30 Thin section of the Cold Bokkeveld carbonaceous chondrite in the petrological microscope, showing a near circular chondrule about 1 millimetre in diameter. The coloured mineral is olivine. Chondrules of high temperature origin are set in a water-rich matrix that is black in this photo. The meteorite contains about 10 per cent of water by weight. The fall occurred in Cape Province, south Africa, on 13 October 1838. A bright fireball and sonic booms preceded a shower of many stones, the largest recovered weighing 2 kilograms.

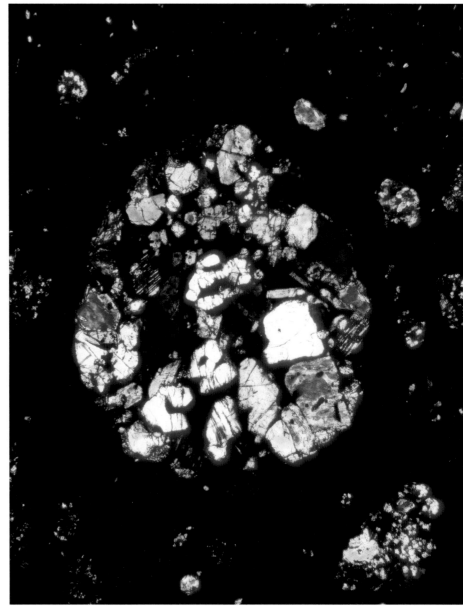

30

STONY METEORITES: ACHONDRITES

Of the 3000 stony meteorites known, about 140 contain no chondrules and formed as igneous rocks (rocks formed by melting such as basalt lava on Earth). Most of the achondrites are fragments of rock that crystallized from molten rock (magma) produced by melting inside an asteroid. There are also twelve stones that came from the Moon and ten (including four observed falls) that probably came from Mars.

Most achondrites are chemically similar to basalts, the most common lavas produced by volcanoes on Earth. Basalts often appear black and featureless because they are made of crystals too small to be seen by the naked eye. If the basalt solidifies within the Earth the crystals are larger because cooling is slower and they have time to grow. The crystal sizes in the achondrites suggest that they cooled fairly slowly, inside a planet or asteroid.

In general, achondrites are thought to be the product of melting on asteroids or planets that had compositions like those of the chondrites. They show chemical variation, for example, there are carbon-rich achondrites which are related to some carbonaceous chondrites.

Soon after formation, planets and asteroids were heated from within and partially melted. This process still produces volcanoes on Earth today but ended 4400 million years ago on the asteroids, 2900 million years ago on the Moon and perhaps 1000 million years ago on Mars.

Heating of the primordial mixture of stony minerals, metal and sulphide produced liquids. Dense, metal-rich liquids sank to become planetary or asteroidal cores. Stony liquids rose and solidified to become basaltic rocks, enriched in calcium and aluminium. The stony material left behind was

31a

31a A basaltic (eucrite) meteorite from the shower that fell at Stannern, Czechoslovakia, in 1808 (6 centimetres across). Fragments of rock with different grain-sizes can be seen in the grey interior. Fragmentation was the result of impacts on the surface of an asteroid. The fusion crust on basaltic achondrites (formed in flight through the Earth's atmosphere) is typically shiny black.

WHAT ARE METEORITES MADE OF?

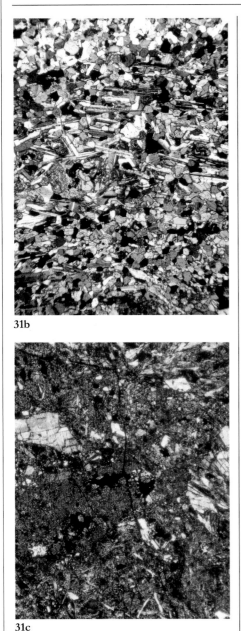

31b

31c

31b Thin-section of basalt lava from the Isle of Skye, Scotland, viewed under the microscope with polarized light (4.5 millimetres across). It is composed mainly of two minerals, both of which are found in basaltic achondrites: Plagioclase feldspar, a calcium-aluminium silicate, is striped black, grey and white (the crystals are aligned in the direction in which the lava flowed); augite, a calcium-magnesium-iron silicate, is coloured. Unlike the meteorite, this basalt has not been fragmented by impact.

31c Thin-section of the Stannern basaltic meteorite, which fell in Czechoslovakia in 1808, viewed under the microscope in polarized light (width of field 4 millimetres). The white mineral grains striped with black are of plagioclase feldspar, a calcium-aluminium silicate. Rarer, more brightly coloured grains are of augite, a calcium-iron-magnesium silicate. Together, these minerals are typical of basalts. Note the fragmented nature of the rock compared with that shown in Fig. 31b.

32

32 Cut surface of the Johnstown, Colorado, hypersthene achondrite (10 centimetres wide).

33 The Kapoeta howardite (6 centimetres high), that fell in the Sudan in 1942. This 'soil' from the surface of an asteroid is a mixture of fine-grained, compacted dust and basaltic rock fragments. Part has been darkened (arrowed) by bombardment by particles from the solar wind.

33

impoverished in calcium and aluminium but enriched in magnesium which occurs in the mineral, olivine. Both solidified liquids and their residues are represented in meteorites by the basaltic achondrites.

Some basaltic achondrites, called **howardites**, are pieces of consolidated 'soil' from the surface of an asteroid. They consist of broken crystals and fragments of basaltic rocks with rare chondritic material. The textures of howardites testify to their compaction and fragmentation by impacts over hundreds of millions of years.

The howardites came from the surface of a body that had no atmosphere. This is known because grains within them have been bombarded by the solar wind. The Sun emits more than just light. Such is its internal energy that it ejects a minute proportion of its matter with great force and the particles stream outwards as the solar wind. Most particles are stopped by our atmosphere, which acts as a protective layer, so rocks at the surface of Earth are hardly affected by the solar wind. The howardites, however, contain crystals that were severely damaged by particles from the Sun so they must have come from the surface of a body with no atmosphere. Damage by the solar wind was predicted to be present in grains in the 'soil' on the surface of the Moon and was found in a howardite on Earth only months before the first Apollo landing in 1969.

Some howardites are 3500 or 4000 million years old and the bombardment by the solar wind occurred at this time. It indicates that the inner Solar System, including the Earth, had an ancient radiation environment similar to that of today. It also suggests that the Sun has not changed much during the intervening period.

WHAT ARE METEORITES MADE OF?

ACHONDRITES FROM THE MOON

The Moon, our nearest neighbour, is the only natural satellite of the Earth. It is unique in the Solar System because of its large size in relation to its associated planet. For example, Jupiter is some 320 times more massive than the Earth, but the largest of its moons, Ganymede, is less massive than the Moon. Seen from the Earth, the full Moon has light and dark areas (Fig. 34). In the past it was thought, wrongly, that the dark areas were covered by water, so they were named seas, or **maria**. The lighter areas were termed **highlands** or **terrae**. It was later observed that no clouds covered the Moon's surface. This showed that there is no atmosphere, and without an atmosphere there could be no liquid water. Despite this, 'mare' is still used for the dark areas of the surface, hence names such as Mare Imbrium, Mare Tranquillitatis and Mare Crisium.

The Moon was the first extraterrestrial object to be visited from the Earth, first by automated spacecraft and later, in July 1969, by man. The six Apollo missions to the Moon (1969-72) returned with a total of 380kg of soil and rocks. The three unmanned Russian space craft, Luna 16, 20 and 24, also brought back samples of lunar soil. Mare and highland regions on the near side of the Moon have been sampled and we now know that it has neither atmosphere nor water and that it has always been dry. Rocks from the mare regions are dark basalts (volcanic rocks). In contrast, many of those from the highlands are **anorthosites**, light-grey rocks containing mainly feldspar, a light-coloured mineral containing calcium, aluminium and silicon.

The lunar samples were formed between 4400 and 2900 million years ago, so are older than most terrestrial rocks. The highland anorthosites are older than the mare basalts. These are

34 Full Moon clearly showing the large darker *mare* regions and the lighter *highland* areas. The surface of the moon is scarred by many craters formed by meteorite impacts. The largest of these craters are the mare basins, hence their generally circular form. These basins were later filled with basalt, a dark, volcanic rock. The highland areas are composed mainly of the pale mineral feldspar.

34

between 3700 and 2900 million years old and the youngest represent the final stage of volcanic activity on the Moon.

The surface of the Moon is heavily cratered by meteorite impacts but little identifiable meteoritic material has been found in the returned samples. Because there is no atmosphere to slow down falling objects, even the tiniest micrometeorites strike the Moon at cosmic speeds. All impacting objects, regardless of their size, are vaporized by the energy of the collision and so there is little meteoritic material preserved in the lunar soil. We now know that most of the cratering occurred over 4000 million years ago during a period of intense bombardment.

Speculation about the possible existence of life on the Moon ended when the lunar samples were studied. No evidence whatsoever was found for

35 The exploration of the moon by Apollo 17 mission. The Apollo programme of lunar exploration ended in 1972 with mission to the highlands. The space craft carried a wheeled vehicle for transporting the astronauts over the lunar surface, which increased the area that could be studied. In this view, Harrison Schmitt is working by a large boulder that has been excavated from below the surface by meteorite impact.

35

WHAT ARE METEORITES MADE OF?

the presence of any living organism, either now or in the past. The Moon is a dead satellite and was never host to life in any form.

In 1979 a meteorite from the Moon was found in Antarctica by a Japanese field party. This remarkable discovery was confirmed by the subsequent recovery of ten more lunar meteorites from various parts of Antarctica. In addition a twelfth stone from the Moon was recently found in Western Australia. It is only because we had a good reference collection of lunar rocks that we can be sure that some of these meteorites came from a mare region and the others came from the lunar highlands. It is clear that some impacts on the Moon were violent enough to eject material from its

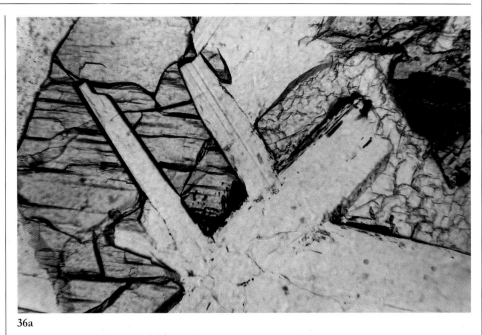

36a

36a A thin section of a lunar mare basalt (1 millimetre across). This volcanic rock was returned to Earth by the Apollo 11 mission. The minerals present are feldspar (white), pyroxene (beige), cristobalite (mottled white) and ilmenite (black).

36b A thin section of a lunar anorthosite (1 millimetre across). This rock was recovered from the lunar highlands and consists of broken fragments of feldspar (white and grey) with small amounts of pyroxene (bright colours). The fragmental texture of this rock shows that it has been shocked by meteorite impact.

36b

gravitational field. Some found its way to Antarctica, taking about eight million years for the journey. During the Apollo missions, each command module measured the composition of the surface rocks as it orbited the Moon. This information shows that the meteorites from the lunar highlands were derived from the far side of the Moon. Thus, at little expense, we have been able to collect samples from part of the Moon not yet visited by mankind.

37 This meteorite (3 centimetres across) was found in 1981 on the ice in Antarctica by an American field party. It is chemically different from most meteorites but similar to rocks from the lunar highlands. It contains angular fragments of feldspar (white).
This is the same mineral seen in the thin section of the lunar anorthosite (Fig. 36b).

37

ACHONDRITES FROM MARS

Eight of the stony meteorites available for study are distinguished from all others by having young formation ages of between 1300 and 200 million years. The eight are collectively known as the SNC meteorites after the names of three of them, **S**hergotty, **N**akhla and **C**hassigny (Back Cover). Evidence from most meteorites indicates that volcanism on their asteroidal parents ended more than 4400 million years ago. Even the Moon, with a diameter of nearly 3500 km was 'dead' by about 2900 million years ago. How then can the eight meteorites be igneous rocks only 1300 to 200 million years old?

It is possible that the eight meteorites formed 1300 to 200 million years ago from the melted material beneath impact craters. If melting had been complete, evidence of an earlier history would have been destroyed. On Earth, such complete melting only

38a

38a Nakhla meteorite. This meteorite fell in Egypt in 1911 as a shower of stones, one of which is said to have killed a dog. This specimen, 12 centimetres across, is covered by a thin, black, glassy fusion crust which formed during passage through the atmosphere.

38b Thin-section of the Nakhla meteorite (4.5 millimetres across). This stony meteorite is composed mainly of interlocking crystals of pyroxene which crystallized from a melt

38b

ACHONDRITES FROM MARS?

occurred beneath very large impact craters many tens of kilometres in diameter. Asteroids would have been torn apart by such intense impacts so the meteorites must have come from a larger body. Alternatively, the SNC meteorites may have crystallized as volcanic rocks between 1300 and 200 million years ago. This parent body must have been larger than the Moon and of planetary size to have had a sufficiently long active life. The Earth, for example, is still active and hot enough to produce lavas from its interior. Having eliminated the Moon and asteroids as the source of the SNC meteorites, the next possibility is Mars. This is supported by the discovery of nitrogen in an SNC meteorite that is heavier than the nitrogen in air but similar to that in the Martian atmosphere, as measured by the Viking spacecraft in 1976.

Clearly meteorites come not only from asteroids, but also from the Moon and from at least one larger planet.

39 Martian surface from the Viking spacecraft. The rubbly surface is characteristic of the surface of Mars. The boulders are up to about one metre across and sit in wind-blown reddish 'soil'.

WHAT ARE METEORITES MADE OF?

Iron Meteorites

As their name implies, iron meteorites – **irons** – are composed mainly of iron-nickel metal, with small amounts of other minerals. Most irons contain 7-15 per cent by weight of nickel. At room temperature, this mixture of iron and nickel does not form a single mineral. On cooling from high

40a

40a Polished surface of the Gibeon iron, etched with acid to reveal the structure (width 14 centimetres). The plates of low-nickel metal forming the Widmanstätten pattern are only 0.3 millimetres thick. They grew in the solid metal when it cooled below about 700°C. The junction between two original crystals of metal is arrowed. The plates of the Widmanstätten pattern have different orientations on either side of the junction. The Gibeon irons were found over a wide area of southern Africa.

40b Polished and etched face of one of the irons of the Sikhote Alin, Siberia, fall of 1947 (width 11 centimetres). The plates of low-nickel metal are 9 millimetres thick. The cross-hatching on each plate is the result of deformation caused by shock from a collision in space. The Widmanstätten pattern is so coarse that it is difficult to see any regularity in the orientation of the plates. The fall in the Sikhote Alin mountains, the largest observed, rained 23 tonnes of fragments over a small area.

40b

temperatures, alloys in this compositional range change to a structural intergrowth of two minerals, one with about 40 per cent nickel and the other with only 5 per cent. This intergrowth, known as the **Widmanstätten** structure, is present in most iron meteorites and formed as

41 Polished surface of the nickel-rich Skookum iron, viewed under a microscope in reflected light (width 1.2 millimetres). This iron has over 17 per cent nickel, too high for the Widmanstätten pattern to form. Instead, tiny spindles of low-nickel metal, with white fringes of high-nickel metal, have grown. The meteorite rusted on Earth to reveal the structure, the dark, rusty outer surface being towards the bottom.

they cooled slowly from a temperature of over 700°C.

Iron meteorites with 5-6 per cent nickel consist almost entirely of the low-nickel mineral, kamacite. At the opposite extreme, those with over 15 per cent nickel contain no visible kamacite. In either case, the meteorites generally appear to be structureless.

Most iron meteorites were originally completely molten and formed in the cores of asteroids. The remainder were not completely melted and formed in metallic pods associated with stony material.

The iron meteorites are classified on a chemical basis. The abundances of nickel and some trace elements were measured in the 500 known irons and when plotted on graphs, 86 per cent of

42 Nickel (Ni) content plotted against gold (Au) and iridium (Ir) contents from chemical analyses of members of two of the thirteen groups of iron meteorites. Members of group IIAB formed in the once molten core of an asteroid. In contrast, members of group IAB were never melted. The different origins are reflected in the variation in the concentrations of the chemical elements. Among the members of group IIAB, as nickel increases from 5.5 to 6.5 per cent, the iridium content decreases by almost 10 000 times, while gold increases threefold. In contrast, in group IAB, iridium shows little decrease and gold remains constant while nickel increases from 6 to 25 per cent. (For explanation, see the text.)

41

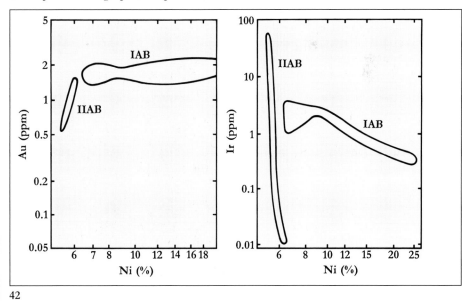

42

the meteorites were found to define thirteen different clusters. Those in each cluster are said to belong to a chemical group of irons and it is thought that each group is from a distict chemical environment – in other words, from thirteen parent asteroids.

Within each group the members have a range of properties. This is illustrated in Fig. 42 for two of the groups. In group IIAB, as nickel increases gold increases but iridium decreases. We know from experiment that when molten iron-nickel metal solidifies slowly, the first solid metal to form is poor in nickel and gold but rich in iron and iridium. As solidification proceeds, the remaining liquid metal becomes enriched in nickel and gold, but poorer in iron and iridium. This causes the chemical variation within the series of solids that form from the liquid, which is what we observe in the irons of group IIAB. We interpret the members of the group as having formed as a series in an asteroidal core as it slowly solidified.

In contrast, the members of group IAB show little variation in gold or iridium although their nickel content ranges from 6 to 25 per cent. This indicates that they were never completely molten and probably formed from small pods of iron-nickel metal distributed within stony material. Indeed, stony material commonly occurs as inclusions within group IAB irons.

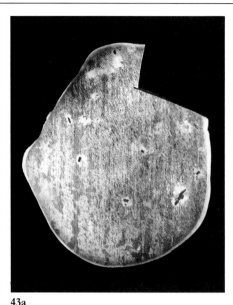

43a

43b

43a Polished and etched surface of the Uwet, Nigeria, iron, of group IIAB (15 centimetres across). The meteorite has only about 6 per cent nickel, so is essentially structureless and composed almost entirely of low nickel metal, with tiny crystals of iron-nickel phosphide.

43b Polished surface of the Landes, group IAB, iron. Note the abundance of dark, angular, silicate fragments, the presence of which testifies that the meteorite was never wholly molten. (Width of slice 18 centimetres.)

STONY-IRON METEORITES

These rare meteorites are of two main types - pallasites and mesosiderites. Pallasites are mixtures of iron-nickel metal and the stony mineral olivine, a magnesium-iron silicate whose gem variety is known as peridot (Fig 44). The olivine grains sit in a metal mesh, sometimes with a continuous Widmänstatten structure. This texture indicates that molten metal was mixed with the olivine crystals at very high temperature and with little damage.

Mesosiderites are composed of chunks of metal, angular fragments of basaltic achondrite, glassy material, and metal veins. Each metal 'slug' developed its own configuration of the Widmanstätten structure, and the presence of glass and metal veins indicates that the whole was subjected to transient and variable high temperatures and fragmentation, followed by slow cooling.

Most stony-iron meteorites are related to the basaltic achondrites and to one group of irons that are inferred to have come from the core of an asteroid. The stony-irons are likely to be from a large layered asteroid such as Vesta (diameter 500 kilometres).

Pallasites formed at a late stage in the melting process that caused metal to sink to form the molten core. We think that cooling, contraction and cracking in the overlying olivine-rich stony mantle led to the injection of molten metal upwards along the cracks.

44

44 Polished and acid-etched slice of the Thiel Mountains, Antarctica, pallasite (6 centimetres across). This stony-iron meteorite is composed of clusters of rounded, pale green olivine crystals set in a mesh of chemically zoned, iron-nickel metal. Such meteorites probably were broken from a layered asteroid at the boundary between its metallic core and olivine mantle. Although iron-nickel metal and olivine occur individually as minerals on the surface of the Earth, intergrowths like this may occur deep within the interior, being totally inaccessible, between the core and mantle.

STONY-IRON METEORITES

There the metal solidified to a mesh that encloses olivine crystals, singly or in broken groups.

From their textures, mesosiderites must have formed differently from the pallasites. Impact must have been involved. Metal slugs excavated from the core were mixed with olivine from the mantle and fragments of basaltic rock from the crust. The various components were welded together to produce the mesosiderites.

45a Polished slice of the Mincy meteorite (37 centimetres across). This stony-iron, from Missouri, USA, is a mesosiderite that formed by the shock mixing of metal and stony fragments. Note the bright metal 'slugs' and the dark, often angular stony fragments. The large central stony fragment includes a metal vein (arrowed).

45b Stony material in the Estherville mesosiderite, viewed down the microscope under polarized light (width of field: 4 millimetres). Feldspar is striped white, grey and yellow. Most of the area is composed of howardite material, that is, fragmented basaltic "soil" from near the surface of the parent asteroid. On the right is part of a centimetre-sized fragmented olivine crystal. The fragments appear mauve, green and orange. The olivine was probably excavated during impact from the interior of the asteroid. This meteorite fell in Iowa in 1879 after the appearance of a brilliant fireball followed by sonic booms.

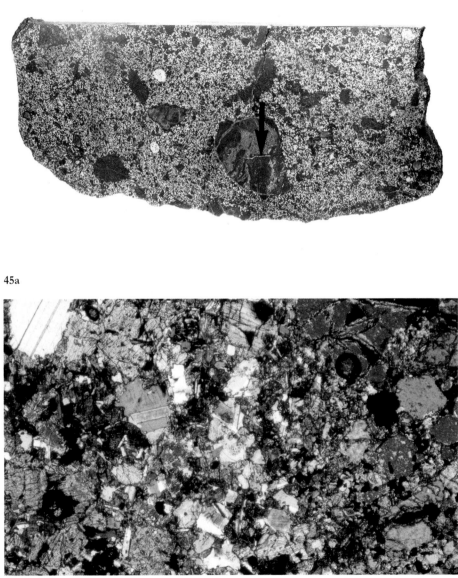

45a

45b

The falls of the different types of meteorites that we observe today indicate that mixing of planetary materials continues. The Earth is still receiving fragments from a variety of sources.

Evidence for the mixing of planetary material in the distant past is obtained from the study of ancient lunar 'soils', which have a measurable meteorite component, and from the occurrence of fragments of meteorite material within meteorites of a different type. For instance, the Barwell meteorite, which fell as a shower of stones in Leicestershire, England, on Christmas Eve 1965, is a chondrite of the type most commonly seen to fall. It has an even grained appearance except for the presence of rare pebbles that were formed by melting on an asteroid of different type.

We now have some twenty-five examples of meteorites within meteorites. These show that fragmentation of asteroids is a continuing process that began during, or soon after, their formation some 4550 million years ago.

46 Cut interior surface of one of the fragments of the Barwell, Leicestershire, fall of Christmas Eve, 1965. The oval object, 15 by 12 millimetres across, is a piece of basaltic rock from a different asteroid from that of the even-grained, grey chondrite in which it now sits. Both pebble and grey chondrite have been together for over 4500 million years.

47 Cut surface of the St Mesmin chondrite that fell in France in 1866. The light coloured chondrite, rich in oxidized iron but poor in iron-nickel metal, contains an inclusion of metal-rich dark chondrite. The dark chondrite has an age of 1300 million years, which may be the time of mixing of the two types. The inclusion is 4 centimetres long.

48 The Bencubbin, Western Australia, stony-iron meteorite (7 centimetres across). This is a complex mixture of iron, one type of achondrite and two types of chondrite. The mixture suffered two or more impacts. The first occurred close to 4500 million years ago and caused partial melting by heating to a temperature of over 1100°C.

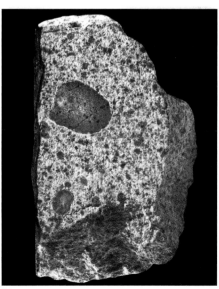

46

47

48

HOW OLD ARE METEORITES?

There are four distinct periods of time that are significant in the history of a meteorite:

1 The terrestrial age – the time spent on Earth;

2 The cosmic-ray exposure age – the time spent as an approximately metre-sized meteoroid in orbit around the Sun;

3 The formation age – the time since the whole meteorite was last heated above about 700°C;

4 The formation interval – the time between the formation of chemical elements in stars and their incorporation into the parent of the meteorite.

1 THE TERRESTRIAL AGE

This is the time since the meteorite landed on Earth, which obviously is known precisely for observed meteorite falls. It is less well known for meteorite finds. When in orbit round the Sun, a meteoroid is bombarded by cosmic rays. After landing on Earth the meteorite is protected by the atmosphere and the unstable products of cosmic radiation begin to decay. We know from observed meteorite falls the normal quantity of these products in a meteorite when it arrives. A meteorite find will have a smaller amount of these products, and the difference depends on the time since the meteorite fell. Carbon-14 'dating' is one of the techniques that may be used to determine the terrestrial age of a meteorite.

Terrestrial ages of meteorites generally range up to a few tens of thousands of years, but many meteorites from Antarctica landed over 500 000 years ago. Only two 'fossil' meteorites are known. (These are meteorites found within rocks on Earth). Both are chondrites from Ordovician limestone in Sweden. Their terrestrial ages are derived from the ages of the fossils present in the sediments that contain them - about 480 and 485 million years.

49 The Brunflo, Sweden, fossil meteorite. This chondrite fell into the sediment beneath a shallow sea over 450 million years ago. It is about 5 centimetres across. Reaction of the chemical elements of the meteorite with the sediment has produced the halo. This occurrence tells us that the meteorites most commonly seen to fall today also fell to Earth during the Ordovician period. Note the fossil – a straight nautiloid, arrowed.

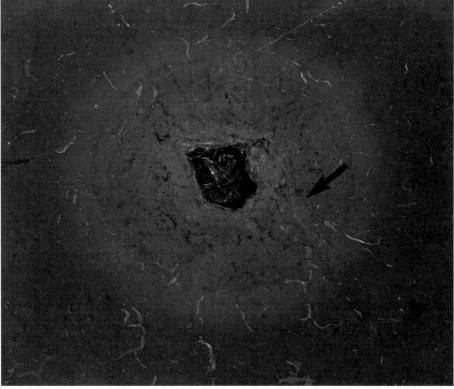

49

This important observation shows that the most common type of meteorite falling to Earth today was also a visitor from space some 500 million years ago.

2 THE COSMIC-RAY EXPOSURE AGE

The second 'age' of a meteorite is the time during which it orbited the Sun as a small body in space. When a rock or lump of iron-nickel metal is in space the cosmic radiation reacts with some of the atoms of which it is formed. These nuclear reactions cause the build-up of secondary atoms. The quantity of secondary atoms produced in this way depends on the chemical composition of the meteoroid and on the duration of its exposure to cosmic rays. Measurements of the abundance

50a One of two stones that fell in 1884 on Tysnes Island, Norway (height of specimen 8 centimetres). The Tysnes Island chondrite has a short cosmic ray exposure age of about 5 million years.

50b Polished and etched slice of the Picacho, New Mexico, iron meteorite. This meteorite and several others of the same group have cosmic ray exposure ages close to 700 million years. This shows that 700 million years ago, an impact event caused fragmentation in the metal core of the parent asteroid. From that time until its fall to Earth, the Picacho meteorite was a small object in orbit around the Sun.

50a

50b

HOW OLD ARE METEORITES?

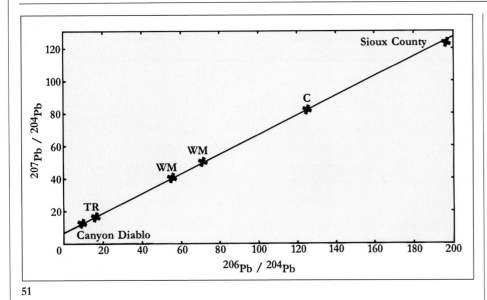

51 This graph shows the determination of the age of three different meteorites. Atoms of the same chemical element may have different atomic weights. These different atoms are known as 'isotopes' of the element. Uranium has two impotant natural isotopes, with weights of 235 and 238 mass units. Uranium-235 decays to lead-207 and uranium-238 to lead-206, but at a much slower rate. If a mineral or meteorite has remained cold then uranium and lead were fixed, and the age can be obtained by measuring only the ratio of lead-207 to lead-206. If the uranium to lead ratio in the meteorite or mineral was high, then more lead-207 and lead-206 would have been produced, and on the graph it would plot high to the right. (The chemical symbol for the element, lead, is Pb, from the Latin, *plumbum*.) Because isotopic ratios are easier to measure than the abundances of isotopes, the two lead isotopes are plotted relative to lead-204, which has no radioactive parent. On the graph are plotted the isotopic ratios of lead extracted from: Barwell 'whole meteorite' (WM) samples, separated Barwell chondrules (C), separated Barwell iron sulphide, the mineral troilite (TR), troilite from the Canyon Diablo iron meteorite, and lead from the Sioux County basaltic achondrite. All lie close to the line whose slope represents an age of 4550 million years. This shows that an iron meteorite and an asteroidal basaltic rock formed at the same time as a chondrite. Furthermore, lead from some new volcanic rocks on Earth plots close to point TR, which is consistent with the Earth's age being about 4550 million years.

of neon gas indicate that stony meteorites have cosmic ray exposure ages of a few millions to a few tens of millions of years. It appears that few stony meteoroids survive collisional destruction by pulverization in space for more than about 40 million years. Iron meteoroids, on the other hand, are much tougher and appropriate measurements show that some have survived in space as metre-sized objects for about 1000 million years.

3 The formation age

This is the time between the present and the last major high-temperature event in the history of the meteorite. The formation age of the basaltic achondrites, for example, is the length of time since they crystallized from a liquid and cooled. Although they remained unmelted, the chondrites were hot and recrystallized as solids soon after they formed. Their formation age is the time since recrystallization, when their constituent mineral grains were produced. Formation ages of both types of meteorite are close to 4550 million years. A very simplified explanation of the technique used to determine formation ages is as follows:

Use is made of radioactive 'clocks' such as the natural decay of uranium to form lead, which occurs at a known rate. We use samples from either a number of related meteorites or the separated grains of different minerals from a single meteorite. In either case the quantities of uranium and lead are determined in each sample, and from these the proportion of lead formed from the decay of uranium is calculated. From this the time since the meteorites were hot can be calculated - that is, the time since uranium and lead were last able to move about freely from one mineral grain to another, or from one type of rocky meteorite to another.

52 Tracks (seen as holes) from the fission of plutonium in a pyroxene crystal from the Tuxtuac, Mexico, chondrite, as seen in a scanning electron microscope. When an atom of an unstable heavy element, such as plutonium, splits by fission, the fragments fly apart with great energy. As a fragment ploughs through a crystal it damages the crystalline structure along a cylindrical trail. When the crystal is etched, the trail corrodes more readily than the undamaged crystal and a fission-track is revealed. The scale bar is 10 micrometres, or 0.01 millimetres.

53a Basaltic rock fragment, 1 millimetre across, in the Semarkona chondrite. From its texture and composition, this tiny piece of rock formed from a melt on an asteroid. The rock is composed of olivine (orange), pyroxene and plagioclase feldspar, (barred white and black). Some of the aluminium was radioactive, and decayed to magnesium.

52

53a

HOW OLD ARE METEORITES?

4 FORMATION INTERVAL

Almost all the elements, other than hydrogen and helium, were formed in stars of various types. This is true not only for meteorites but for everything on Earth, including our bodies. The formation interval of an element is the time between its formation in a star and its incorporation into the material that became a planet or meteorite.

Many stony meteorites contain the fission products of plutonium (see Fig. 52). Plutonium is an unstable element which decays very quickly — its half-life is only 82 million years, as opposed to uranium-238 which has a half-life of 4500 million years. (The half-life of an element is the time taken for half of the atoms in any mass of the element to decay into other elements.) Because of its short half-life, all the plutonium originally present at the formation of the Solar System had decayed by about 4000 million years ago — plutonium no longer occurs naturally on Earth or anywhere else in the Solar System. Measurements of the products of plutonium decay in meteorites have shown that the formation interval for plutonium was about 150 million years. This means that plutonium was formed in a star just 150 million years before the formation of the Sun and planets.

A proportion of some other chemical elements formed much closer to the time of planet formation. New evidence indicates that some planetary bodies were melted by a type of radioactive aluminium. This aluminium is highly unstable and must have formed in a star less than five million years before the planets were born. So, different chemical elements enable us to identify stages in the formation of the embryonic Solar System.

The radioactivity of meteorites is much lower than in Earth rocks, which are richer in uranium and thorium. Extremely sensitive equipment is required to measure radioactivity in meteorites.

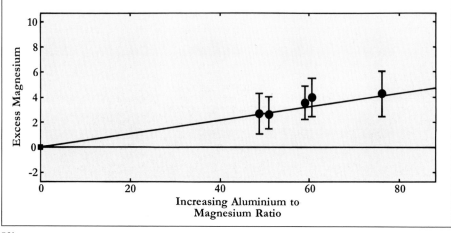

53b

53b Magnesium from the decay of aluminium (vertical axis) plotted against the aluminium to magnesium ratio in the minerals of the rock in Fig. 53a. When the Solar System formed, a highly unstable, radioactive isotope of aluminium was present but within a few million years it had completely decayed to magnesium. On the plot, the square at zero represents olivine and pyroxene crystals that are poor in aluminium and rich in magnesium. The magnesium from the decay of radioactive aluminium is not significant compared with the amount of magnesium originally present – there is no measurable excess. Crystals of plagioclase feldspar have high aluminium to magnesium ratios, so the magnesium from the decay of radioactive aluminium is measurable (the five points, with error bars, on the right). The slope of the line indicates that when the rock cooled, each million atoms of aluminium had eight that were radioactive. From the quantity of aluminium in different types of meteorites there would have been enough radioactive aluminium to have partially melted asteroids 100 kilometres in diameter. Radioactive aluminium was a potent heat-source in the early Solar System and its formation interval was only about 5 million years.

Most meteorites are fragments broken from asteroids. Most asteroids, however, orbit the Sun in almost circular, stable orbits in the asteroid belt, between Mars and Jupiter. Meteorites, then, are from strays from the asteroid belt that were diverted by the gravity of Jupiter into elliptical, earth-crossing, orbits.

It is difficult to estimate how many stray asteroids are represented in the world's meteorite collections. Each of the thirteen groups of iron meteorites probably comes from a different asteroid. In addition, there are nine chemically distinct groups of chondritic meteorites each of which may share the same parent with one of the groups of

54 The largest surviving piece of the Farmington meteorite which fell in 1890, in Kansas. The cut face is 43 centimetres long. The veins of shiny metal were produced by intense shock in space.

54

METEORITES & ASTEROIDS

irons. This is certainly true for some pairs of chondrite and iron groups. Some individual chondrites and dozens of irons are mineralogically or chemically unique, and may be rare representatives of other asteroids. So we can conclude that at least thirteen, but perhaps over fifty, asteroids have been the source of the meteorites available to science.

One recent attempt to identify the asteroid from which a meteorite came is particularly interesting. The Farmington, Kansas, meteorite that fell on 25 June 1890 proved to have an exceptionally short cosmic ray exposure age of about 20000 years. This is so short that the orbit probably did not change after the meteoroid was broken from its parent asteroid. Therefore, determination of the orbit of the meteorite would give that of the parent, which should then be identifiable from observation. Eye-witness accounts of the path of the fireball were gleaned from contemporary local newspapers by the Historical Societies of Kansas and Nebraska and were used to determine the trajectory of the meteoroid in space. One 98 year old (in the mid 1970s), however, was able to tell his own story. Sixty reports from 25 locations in and around the place of fall were selected by the scientists for their determination.

Although the meteorite fell at 12.50 on a midsummer day, 'It was a wonderful sight and the light was fully as strong as the sun itself'. The scientists deduced that the meteoroid probably came from 20 degrees west of south, and descended at 60 degrees from the horizontal. It is impossible to estimate the velocity from visual observations, so different orbits were calculated using different velocities between 13 and 22 kilometres per second. It was concluded that the parent asteroid may have been 1862 Apollo, Hermes or 1865 Cerberus, with two other less likely possibilities.

This account is a good illustration of the dependence of scientists on the non-scientific community for observations of rare, transient events such as meteorite falls.

The study of meteorites provides information on the origin and history of their parents, the asteroids. For example, many iron meteorites appear to have come from the iron-nickel cores of asteroids and contain evidence of the time of melting, of formation of the cores and the rates at which they cooled.

Small asteroids of less than 100 kilometres in diameter may not have melted, so they probably do not have cores. But within a few million years of their formation, large asteroids melted to produce cores. We know this from measurements of the abundances of palladium and silver in the minerals of iron meteorites. Palladium is a **noble metal**, like gold, which means that it is chemically unreactive and tends to stay as metal rather than form stony minerals. Some palladium was formed

55a

in an exploding star, or **supernova**, debris from which was incorporated into the gas and dust just before the planets began to form. A fraction of the palladium was radioactive and decayed to silver within 50 or 60 million years.

In iron meteorites, palladium is contained in the iron-nickel metal, but silver is concentrated in iron sulphide. By measuring the amounts of palladium and silver in both metal and sulphide, the amount of silver produced by the radioactive decay of palladium in the metal can be calculated. Results indicate that the asteroids formed, melted to produce cores, and the cores began to cool within 20 or 30 million years of the original supernova event.

Thus, from the study of meteorites we can determine the time-scale of planetary formation and melting. The Solar System is about 4550 million years old, but the asteroids, and presumably the planets like our own, formed and melted at the very beginning of Solar System history.

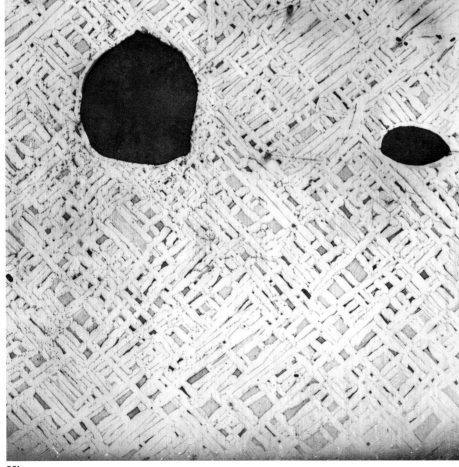

55b

55a A 20 tonne mass of the Cape York, west Greenland, iron meteorite shower, when found by Dr Vagn Fabritius Buchwald. The meteorite is the dark object, with hammer, at lower left. Because the fjord is free of ice for only a few weeks in the year, recovery of the meteorite, now in Copenhagen, was very difficult.

55b Polished and etched part of a slice of the same meteorite, 12 centimetres across. Note the Widmanstätten pattern and the two dark nodules of the iron sulphide, troilite. The iron nickel metal contains excess silver from the decay of radioactive palladium. The troilite contains more silver, but no measurable excess from palladium decay. The meteorite belongs to a group that formed by the solidification of a once molten asteroidal core. The presence of a measurable amount of silver from the decay of radioactive palladium indicates that melting and solidification occurred within 20 or 30 million years after the production of the palladium.

THE SIZE OF PARENT ASTEROIDS

We can estimate the approximate size of an asteroid from the rate of cooling of its core. Large objects take longer to cool than small ones. The rate at which an iron meteorite cooled can usually be determined from its nickel content and the coarseness of its internal structure. On cooling, alloys of iron-nickel metal with between 7 and 15 per cent nickel form a structural intergrowth of bands of metal with 5 per cent nickel set in metal with about 40 per cent nickel (the Widmanstätten structure, see p. 35). The thickness of the bands of nickel-poor metal is controlled by the nickel content of the whole meteorite and by the rate at which it cooled below about 700°C. If the nickel content is high, then the nickel-rich matrix is abundant and the bands of nickel-poor (5 percent) metal are thin. If cooling is fast, the bands of nickel-poor metal have little time to form, and this also causes them to be thin. Conversely, an iron meteorite with only 7 per cent nickel, if slowly cooled, will have a coarse structure with thick nickel-poor bands.

From 600 to 400°C many irons cooled at rates between 10 and 100 °C per million years. The parent asteroids of most iron meteorites probably had stony shells around the metal cores, which would have cut the rate of heat-loss. The measured range of cooling rates indicates that these asteroids had overall diameters of a few hundred kilometres, like some asteroids observed by astronomers today.

This evidence corroborates our earlier conclusion that most meteorites come from asteroids. Furthermore, it is an indication of how widespread melting must have been among even the small planetary bodies of the early Solar System

56a & b Polished and etched slices of two iron meteorites having the same nickel content, 7.4 weight per cent, and the same height, 9.5 centimetres.

56a The Jamestown, North Dakota, meteorite has a fine structure, the bands being only 0.3 millimetres thick.

56b The Bohumilitz, Czechoslovakia, meteorite has a coarse structure, the bands being 1.9 millimetres thick. This meteorite cooled more slowly than Jamestown; the bands in Jamestown had little time in which to develop.

The Jamestown meteorite is notable for its deformed structure which is similar to that of the iron which produced Meteor Crater in Arizona, USA. Some of the bands are curved, due to shock deformation by impact on an asteroid. The meteorite must have been excavated from the asteroid during a second impact, and sent on its journey to Earth.

56a

56b

DO SOME METEORITES COME FROM COMETS?

Comets are thought to be composed of grains of stony material bound together by a variety of ices to form a 'dirty snowball'. This is based on observations of comets when they are near the Sun. The Sun's radiation causes the surface of a comet to evaporate, which releases stony particles to form a dust-tail. Dust-tails may be distinct from the glowing, electrically charged atoms that produce the spectacular cometary tails sometimes visible from Earth.

The mass of ices and stony material at the core of a comet is known as the nucleus. Until the recent passage of Halley's Comet, no cometary nucleus had been 'seen'. This is because the nucleus of a comet is so small that it generally cannot be detected by telescope when the comet is beyond Jupiter. Comets only become visible when they release glowing material as they approach the Sun, and this material shields the nucleus from view. But in 1986, the spacecraft *Giotto* was sent close enough to Halley's Comet to take images of the nucleus (Fig. 58).

A few comets travel around the Sun on orbits like those of Earth-crossing asteroids. These **short-period** comets differ from asteroids in having a fuzzy appearance. Other, 'periodic' comets, such as Halley, travel on more elliptical orbits that take them beyond the outer planets. Most, however, appear to come from the outermost reaches of the Solar System, far beyond the orbits of Pluto and Neptune. These are **long-period**

57

comets that take some millions of years to orbit the Sun. Because none has been seen before, each arrival in the inner Solar System is greeted as a 'new' comet. None appears to be from beyond the Solar System, and the inference is that, to account for the arrival of several 'new' comets each year, they must come from a cometary 'cloud' comprising millions of comets. The presence of such a cloud was postulated by the astronomer, Jan Oort, so it is known as the 'Oort cloud', in his honour. It must be emphasised that the Oort cloud has not been seen; its presence is only inferred from the motions of these long-period comets.

The origin of comets is still disputed. Astronomers believe that they

57 The head and tail of Halley's Comet, photographed in March 1986. The kinks in the lower part of the tail result from the interaction of electrically charged particles with the magnetic field produced by particles of the solar wind. Dust particles are not electrically charged, and form the upper, straight portion of the tail.

formed by the aggregation of dust, with the condensation of gases to ices, as a cement. This could have occurred within a proto-solar nebula, or interstellar cloud from which the Sun and planets formed. Another possibility is that the comets formed in interstellar clouds encountered by the Solar System on its travels through the heavens, and

DO SOME METEORITES COME FROM COMETS?

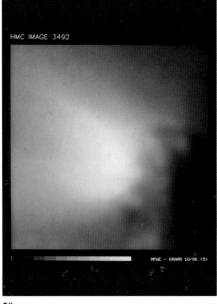

58

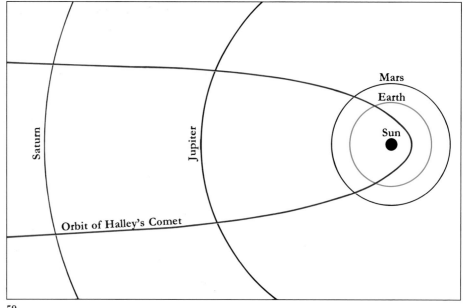

59

were captured by it. Perhaps the most popular theory is that comets formed in the region of the outer planets. After their formation, the comets were swung into highly elliptical orbits that caused them to spend most of their time in the outermost part of the Solar System, in the Oort cloud.

Whatever their origin, there is general agreement that comets are made of some of the most primitive and least altered matter of the Solar System.

It is most important for us to obtain samples of comets. Radar observations of the trails of meteors in the atmosphere show that most have cometary orbits, which in turn indicates that much of the dust entering our

atmosphere is cometary. About half the micrometeorites collected by aircraft in the stratosphere (see p. 4) contain water and carbon dioxide. Such interplanetary dust is composed of minerals like those in several rare groups of stony meteorites, so it is possible that all come from comets. This is largely confirmed by the visible light emitted by shooting stars. The spectra obtained are consistent with the derivation of meteors from water-bearing meteorite parents. Meteors, however, tend to break up in atmospheric flight, which indicates that they are composed of more friable material than the meteorites to which they may be related.

Some carbonaceous chondrites

58 Image of the nucleus of Halley's Comet sent by the Giotto craft when it was 5000 kilometres from the comet. Note the bright jets coming from the comet's interior. The Sun's energy had driven off the ices from the surface, which is dark, but evaporation of ices in the interior was continuing, to form the jets. The nucleus is over 10 kilometres across.

59 The Solar System showing the inner part of the orbit of Halley's Comet, relative to the orbits of Saturn and other planets. As Halley's comet approaches the Sun it speeds up, but slows as it retreats. It spends some four-fifths of its time beyond Saturn.

contain veins in which the minerals were deposited by liquid water. It is a matter for speculation whether liquid water could have existed in a cometary nucleus or on a small asteroid. This is because heating in small bodies with low gravity would cause water ice to evaporate, that is, the ice would become a gas without going through a liquid water stage. But it does seem possible that heating and loss of water from the surface of a small asteroid or comet could have produced a dry skin that could have contained water at pressure inside. This could have allowed liquid water to form and percolate through cracks in the interior, and is consistent with the observations of veins in some meteorites.

It is therefore possible that we may already have pieces of cometary material on Earth. Water-bearing micrometeorites certainly may have come from comets and some water-bearing meteorites may have the same parentage. What we do lack, however, is a chunk of comet complete with its primordial ices, which would give us information on the material from which the outer planets were largely built. Return to Earth of such a sample must await a space mission to a comet, a challenge shortly to be met by the Rosetta mission. The European Space Agency plans to send the Rosetta craft to a short-period comet to analyse its surface.

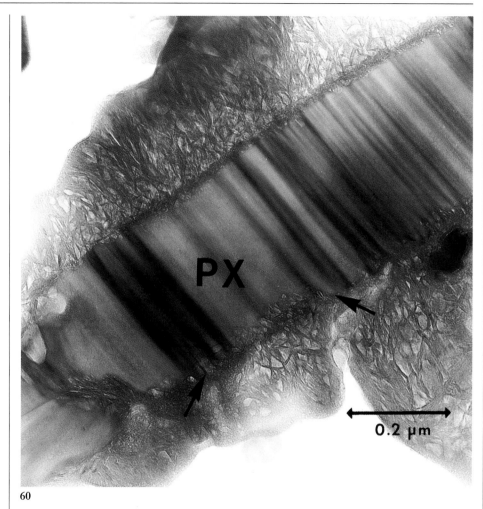

60

60 A specially prepared thin section of the Semarkona chondrite (fron India), showing a fragment of pyroxene (PX) partly altered to clay, which contains water. Between the arrows the curved fibres of clay are clearly eating into the pyroxene. Pyroxene is a magnesium-iron silicate that formed at high-temperature. The clay formed later, at low temperature. The photograph was taken in a transmission electron microscope at the University of Essex. The scale bar is two ten-thousandths of a millimetre.

TIME & THE ORIGIN OF PLANETS

Astronomers observe stars forming in clusters within dense clouds of gas and dust that may be thousands of times more massive than the Sun. The Sun is a star, so the Solar System probably was born in a similar environment.

Chondrite meteorites may be largely unaltered left-overs of the gas and dust from which the planets formed. Alternatively, even if the chondrites resulted from planetary collisions, some contain rare mineral grains from the atmospheres of stars. Furthermore, the 'solar' composition of chondrites (see p. 21) suggests that their origin is intricately bound to the formation of the Sun and planets, and therefore to ourselves. However, there are several conflicting views on the origin of planets and there is no consensus on how the chondrites came into being.

Within the past few years new evidence from meteorites has provided some detail on the earliest part of Solar System history (Fig 63). The interstellar cloud of dust and gas in which the Sun formed probably existed more than 4700 million years ago. At about that time, the plutonium that became incorporated into meteorites was made in an exploding star or **supernova** within the cloud. Afterwards, the existence of the cloud and the formation of other new stars must have continued.

We know that matter from which the planets formed was processed in about ten different stars, as was the matter that makes us! The evidence for this stems from the presence of small amounts of tiny diamonds and other

61 The Horsehead nebula in Orion, a dust-rich region in which stars are forming. The picture was achieved by superimposing images using three different colour filters. The red colour is produced by bright young stars behind the dust. The dust absorbs their energetic blue light and emits red light. The two bluish nebulae below the Horsehead are produced by bright stars in front of the dust, which reflects their blue light.

62 Star-dust in the laboratory! Diamond grains from the Allende meteorite (Mexico), magnified 80 000 times in a scanning electron microscope. Each diamond 'grain' is a plate about forty millionths of a millimetre across. The plates were probably deposited by gas as it streamed out from a star. The plates formed as a coating on other mineral grains. These diamonds are older than the Sun and planets – they were formed before the Solar System.

61

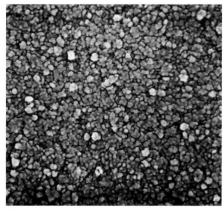

62

TIME & THE ORIGIN OF PLANETS

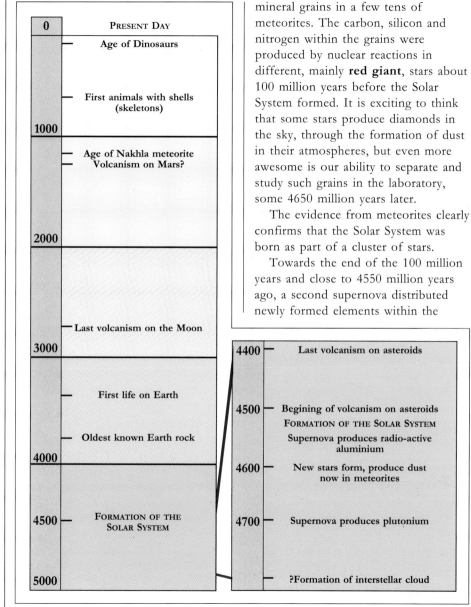

0	PRESENT DAY
	— Age of Dinosaurs
	— First animals with shells (skeletons)
1000	
	— Age of Nakhla meteorite Volcanism on Mars?
2000	
	— Last volcanism on the Moon
3000	
	— First life on Earth
	— Oldest known Earth rock
4000	
4500	— FORMATION OF THE SOLAR SYSTEM
5000	

4400	— Last volcanism on asteroids
4500	— Begining of volcanism on asteroids FORMATION OF THE SOLAR SYSTEM Supernova produces radio-active aluminium
4600	— New stars form, produce dust now in meteorites
4700	— Supernova produces plutonium
	— ?Formation of interstellar cloud

mineral grains in a few tens of meteorites. The carbon, silicon and nitrogen within the grains were produced by nuclear reactions in different, mainly **red giant**, stars about 100 million years before the Solar System formed. It is exciting to think that some stars produce diamonds in the sky, through the formation of dust in their atmospheres, but even more awesome is our ability to separate and study such grains in the laboratory, some 4650 million years later.

The evidence from meteorites clearly confirms that the Solar System was born as part of a cluster of stars.

Towards the end of the 100 million years and close to 4550 million years ago, a second supernova distributed newly formed elements within the cloud. After a further few million years the Solar System formed, perhaps triggered by the exploding supernova compressing part of the cloud.

The first planetary objects possibly formed, became hot and melted before the interstellar cloud had completely dispersed. This is inferred from the presence in a few meteorites of both star-dust and fragments of igneous rock from planets which may have been only 50 or 100 kilometres in diameter. A few millions of years after the formation of the Solar System, the interstellar cloud broke up and the stars moved apart, leaving the Sun and planets together, but alone, in space.

From meteorites we know that asteroidal sized bodies were heated to melting soon after their formation and it seems inevitable that the Earth, too, melted. The occurrence of meteorites within meteorites indicates that collisions between asteroids caused mixing of different types of material. So, towards the end of its birth, the Earth must have received a contribution of different materials from other parts of the Solar System. Early heating of the Earth

63 History of the Earth related to the Solar System. Note that the time since the dinosaurs became extinct is only 1.5 per cent of the age of the Solar System. Information on events before 4400 million years ago – the expanded part of the figure – can only be obtained from meteorites.

63

TIME & THE ORIGIN OF PLANETS

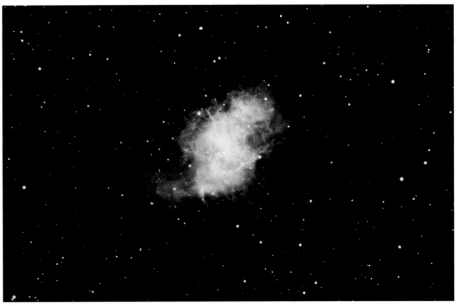

64

64 The Crab nebula, the remains of an exploding star, or supernova, that was observed in 1054 AD by Chinese astronomers. The matter processed in the star is still expanding outwards into space at some thousands of kilometres per second.

would have caused the core to form and grow, which would have released even more heat. Any water or primitive atmosphere probably was lost to space. The early Earth may have been dry, like the Moon today. The oceans and atmosphere were probably formed from the gas and water released by later bombardment of cometary bodies.

Until we can obtain samples of comets and asteroids from space missions, meteorites and their diamonds from the stars will continue to be our only source of information on our distant origin. Hence they provide the key to our understanding of the crucial first links in the chain of events that led to our existence.

THE EARTH

How the Earth and its planetary neighbours formed is uncertain. As we have seen, one theory suggests that the planets may have grown from dust condensed from a hot nebular gas with the composition of the Sun. As the hot gas cooled, different mineral grains formed. Grains that condensed at the highest temperatures, about 1500°C, were rich in the elements calcium and aluminium. In fact, calcium and aluminium-rich inclusions occur in some chondrites and have been interpreted as representing the earliest condensed solids of the Solar System. During the next stage of condensation, iron-nickel metal formed in the inner part of the nebula, near the Sun, where the pressure was highest, but stony minerals condensed further out. The clumping together of metal grains may have led to the formation of Mercury, Venus, Earth and Mars core-first, followed by their stony mantles. Thus, as the gas cooled and condensed, the composition of the growing planets changed, until water was added to those further from the Sun.

A less popular theory argues that, at first, only giant planets, like Jupiter, could have formed. The orbits of the original planets were unstable, which resulted in collisions between them. Collisional break-up of one giant planet led to the formation of Mercury, Venus, Earth (and Moon) and Mars from pieces of its core. Alternatively, a collision between the Earth and a Mars-sized planet (but not Mars itself) possibly caused matter to be ejected from the young Earth to form the Moon.

Which ever way the planets formed, collision between the earliest bodies was common. The planetary and asteroidal fragments so produced were scattered around the Solar System and caused further catastrophic impacts on the planets and their moons. This early, intense bombardment gradually died out about 3900 million years ago. Our atmosphere and oceans may have been produced by the degassing of cometary or 'wet' asteroidal projectiles that landed on Earth towards the end of the bombardment. Afterwards, the Earth entered a period of stability during which life evolved. The Moon, however, was too small to hold water and other gases, so it remained dry, with no atmosphere and without life.

A hundred and fifty years ago, the geologist Lyell argued that the present is the key to the past, implying that the geological processes of today occurred at the same rate throughout geological time. The evidence presented here indicates that this is not strictly true. Some 4550 million years ago the Earth formed, melted, boiled off any water and lost its original atmosphere. Water and atmosphere were replenished by bombardment of the Earth by asteroids and comets from space. By analogy with the record on the Moon, intense bombardment ended about 3900 million years ago. Thus, the formative events of the Earth's early history were neither continuous nor repetitive.

It is, therefore, more accurate to say that 'the present is the key to the past 3900 million years'.

Our planet has fostered life for most of its history (Fig 63), but animals with shelly outer skeletons or with backbones were absent during the first 4000 million years. The Moon has been inactive for the past 2900 million years,

but active volcanoes probably existed much more recently on Mars and they still exist on Earth.

There is no evidence of the early bombardment on Earth because the crust has constantly changed. However, it is still being bombarded from space, but at a slow rate. Craters from hundreds to thousands of metres in diameter are recognized as being the result of impacts during the past tens of thousands of years. The most recent, large event was an explosion in Siberia, on 30 June 1908 (Fig. 69), which occurred some six kilometres up in the atmosphere. There is evidence that a fragment of comet was responsible.

A crater one kilometre across is formed every 1300 years or so, but larger cratering events are less frequent.

65 Polished surface of the Vigarano carbonaceous chondrite which fell in Italy in 1910. The prominent, white, calcium-aluminium-rich inclusions, up to about one centimetre across, may be the oldest objects that formed in the Solar System. (Note: The diamonds shown in Fig. 62 are thought to have formed before the Solar System.)

66 An inclusion of water-bearing carbonaceous chondrite in dry ordinary chondrite (Supuhee, India). Polished surface viewed in reflected light. (Width of field 2 millimetres.) The inclusion is the fairly uniform grey area in the centre that has been invaded by shock-emplaced metal (white).

65

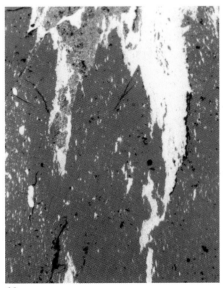

66

TIME & THE ORIGIN OF PLANETS

For example, statistics indicate that a 50 to 100 kilometre wide crater is produced somewhere on the Earth only about once every 100 million years. Such large impacts cause devastation over millions of square kilometres and affect the whole world. Although asteroidal impacts pose a real threat to humanity, the probability is that none will occur in the foreseeable future.

Geological processes have largely worn away the crater structures caused by older impacts – all that are left are scars in the rocks that originally lay beneath the craters. Such ancient impact scars are known as 'astroblemes'. Some **astroblemes** are many tens of kilometres in diameter, and occasionally they occur in pairs, or groups. Astroblemes testify to the occurrence on Earth of major impacts that are random in place and time.

In some cases it is possible to identify the type of meteorite that caused an impact scar. This is because on striking the Earth a large meteoroid explodes and impregnates the underlying rocks with a vapour that carries its chemical signature. On Earth, **noble metals** such as platinum, gold and iridium are concentrated in the core. Compared with most meteorites, rocks of the Earth's crust are highly impoverished in these metals. Thus, the presence of excesses of noble metals in the deformed rocks of circular structures is a reliable indicator that a meteorite was the culprit.

Large impacts cause vaporization and melting of the country rocks. Some of the material is blown out through holes created in the atmosphere and returns to Earth as glassy objects called tektites. For example, **tektites** found in Czechoslovakia have an age of nearly 15 million years. They are thought to have been formed by an impact that produced a 24 kilometre diameter astrobleme of the same age at Nordlinger-Ries, in southern Germany.

67 Meteor Crater (Barringer Crater), Arizona. Meteor Crater is 1.2 kilometres across and 170 metres deep. Note the raised rim that stands above the level plain. This, the world's best preserved large crater, was formed about 50 000 years ago. Some 30 tonnes of iron meteorites have been found on the rim and over the surrounding plains. The bulk of the meteoroid vaporized on impact and was destroyed.

68 The scar of Manicouagan (Quebec, Canada) seen from a satellite. It was produced by an impact 210 million years ago. A lake (dark) forms a ring 65 kilometres in diameter. The crater has been eroded away, but the effects of the underlying deformation are still visible in the circular structure. Shock effects are displayed by the rocks.

67

68

METEORITE IMPACT & EVOLUTION

During the past decade it has been emphasised that the evolution of plants and animals occurred in fits and starts. Over tens or hundreds of millions of years, many plants and animals appear to have changed little, but steadily. Six such periods of steady evolution are known, but each of the first five ended in a biological revolution.

The latest biological revolution is the best known. For 150 million years reptiles dominated the Mesozoic Era (Fig. 63). Then dinosaurs, pleisiosaurs and ichthyosaurs (marine reptiles) and flying reptiles all became extinct, and mammals became dominant. The most complete extinction, however, occurred in the micro-organisms that lived near the surface of the sea. Most forms died out some 65 million years ago, at the end of the Cretaceous period and before the Paleocene, the beginning of the Tertiary era. This extinction was 'sudden', which, in palaeontological terms means within half a million years, and perhaps much less. Other animals, such as birds, crocodiles, turtles and insects, and many plants, were unaffected. After the Mesozoic, in the early Tertiary Era, evolution of the survivors continued, which led to the plants and animals of today.

There has been great speculation and controversy over the causes of the major biological changes during Earth history. Much publicity has been given to the evidence for meteorite impact at the Cretaceous-Tertiary boundary, about 65 million years ago. This evidence is particularly compelling, and includes world-wide enrichment in noble metals in a thin layer at the boundary. The enrichment has been found not only in clay deposited on the sea-floor, but also in sediments laid down on the continents. Often, the noble metals are accompanied by mineral grains showing evidence of intense shock, and, in a few localities, by a layer of soot.

The evidence indicates that one, or more, major meteorite impact occurred at the end of the Cretaceous, and at least one must have been on a continent. (The soot is thought to be the product of ensuing forest fires.) Acid rain would have been produced. It would have acidified the surface waters of the oceans and so dissolved the shells of many marine animals. Soot and dust in the atmosphere would have shut out sunlight and caused cooling for a time, so life would have been very difficult for land animals and plants.

Many palaeontologists, however, are sceptical of the proposed link between meteorite impact and biological extinction and evolution. They cite a lack of firm evidence for impact during the other four major biological crises, such as the Ordovician- Silurian and Permian-Triassic boundaries. Changing climate and conditions in the seas wrought by the movement of the continents are deemed to be more likely causes of enhanced biological change. Volcanism may have been involved, especially at the end of the Cretaceous, when huge eruptions occurred in India. In addition, if impact had caused the change that marks the end of the Cretaceous, there ought to be evidence of adverse effects on land plants, turtles and insects. Absence of this evidence is taken by palaeontologists as favouring a mechanism less cataclysmic than impact and produced by changes on the Earth itself. That a major impact occurred near the end of the Cretaceous seems beyond reasonable doubt, but it need not have been related to biological change and the demise of the dinosaurs.

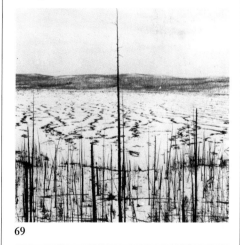
69

69 Dead forest, Tunguska area, Siberia, in 1929. On 30 June 1908 there was an explosion in the atmosphere, probably caused by a small piece of comet. Here the trees still stand, but stripped of branches, indicating that the blast was overhead. Further from the centre the trees were felled, with trunks pointing away from the centre. This is an indication of the destructive power of impact from space.

FUTURE RESEARCH

For two centuries the study of meteorites has gone hand in hand with the development of technologies and theories of our origins. It is no different today. Specially adapted optical microscopes were used to examine thin-slices of meteorites in the 1860s, before the technique became widely used in the study of rocks.

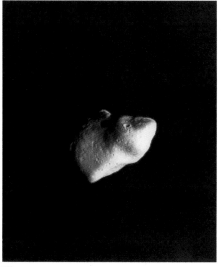

70

70 Asteroid 951 Gaspra in the blackness of space. This first image of an asteroid was taken by the Galileo craft from 16 200 kilometres (10 000 miles) on 29 October 1991. The Sun is to the right. Gaspra was unevenly shaped by impact, being 20 x 12 x 11 kilometres (12.5 x 7.5 x 7 miles) in size. Such remote sensing will be important in future studies of asteroids.

Precise ages were determined for meteorites before they were known for Earth rocks. The primordial objects from space, known as meteorites, clearly fire the imagination of scientists the world over. But what of the future?

Within the next twenty years, samples may be returned to Earth from Mars and from a comet,

71

71 The biggest meteorite found in the MacAlpine Hills, Antarctica, during the 1988-89 field-season. A party of eight recovered 1078 meteorites in six weeks on the ice. Two of the meteorites are from the Moon. This is part of the Ansmet (Antarctic Search for Meteorites) programme. The Antarctic will continue to be a source of new and exciting meteorites for years to come.

and we must be ready to receive them. Our preparations have already begun. A number of scientists are paying particular attention to the SNC meteorites as possible martian samples. Rare clays and calcium carbonate in the meteorites may yield clues to weathering processes on the planet, and the included gases are being compared with the analysis of the martian atmosphere obtained by the Viking missions in 1976. The absence of life on Mars has not yet been conclusively proven, so one problem requiring a solution is the quarantine arrangements for returned material.

The handling of cometary samples will prove to be more difficult. The space missions to Halley's Comet indicate that much of its dust is exceedingly fine. Techniques are already being applied to micrometeorites that enable us to obtain chemical analyses and estimates of their ages. In Germany, scientists have prepared synthetic cometary material to determine the effects upon it of radiation in space. Perhaps the most difficult challenge, however, will be to make the various measurements on cometary material at the temperature of cometary interiors, perhaps -200°C. Otherwise structural information will be lost. But whatever the problems, the challenge will be met.

It is with the continuing study of meteorites and comets that we shall unlock the mysteries of our beginnings.

Ablation Erosion of the surface of a meteoroid by friction in the atmosphere; or of Antarctic ice by wind.

Achondrite Stony meteorite composed of the products of melting on *asteroids*, the Moon or on a planet.

Anorthosite Rock composed of calcium-feldspar.

Apollo asteroid Asteroid that passes inside the Earth's orbit.

Asteroid Small planet or fragment of planet in orbit around the Sun.

Astrobleme Ancient 'scar', formed by meteorite impact.

Ataxite Structureless iron meteorite, rich in nickel.

Basalt Fine grained volcanic rock rich in calcium and magnesium.

Blue ice area Zone of compressed ice rising to the surface.

Carbonaceous chondrite Class of magnesium-rich stony meteorite; many contain water and organic compounds.

Chondrite Stony meteorite not melted since it formed by the aggregation of different components.

Chondrule Object formed from a solidified melt; a component of most stony meteorites.

Comet Body of ice and dust in orbit around the Sun.

Core Central concentration of iron-nickel metal within an asteroid, planet or moon.

Cosmic ray exposure age Time when a meteoroid was exposed to radiation in space, or a rock was exposed on the surface of a body.

Cosmic spherule Droplet from space, usually smaller than one millimetre.

Diogenite Achondrite composed of pyroxene.

Escape velocity Velocity required by an object for it to leave the gravitational field of a body.

Eucrite Class of achondrite with a basaltic composition.

Explosion crater Crater formed by high velocity impact.

Fall Recovered meteorite whose fall was witnessed.

Feldspar Group of silicate (stony) minerals with calcium, sodium or potassium, plus aluminium.

Find Meteorite whose fall was not witnessed.

Fireball Light in the atmosphere associated with a meteoroid.

Formation age Time since a rock was hot enough to cause movement of chemical elements between different minerals.

Formation interval Time between the formation of the atoms of a chemical element and their incorporation into a planetary body.

Fusion crust Layer formed by the freezing of surface melt on a meteorite at the end of high velocity atmospheric flight.

Gravitational focussing Bending of the path of an incoming object towards the centre of gravity of a more massive body.

Hexahedrite Apparently structureless iron meteorite with a low nickel content.

Highlands (lunar) Bright regions on the Moon comprising ancient crust dominated by anorthosite.

Howardite Class of achondrite; surface 'soil' of an asteroid.

Igneous rock Rock formed by the solidification of molten rock.

Individual One of a shower of meteorites from a single fall.

Interplanetary dust Micrometeorites that underwent little heating in atmospheric flight.

Interstellar cloud Region in space where gas and dust are concentrated.

Iron Meteorite dominantly composed of iron-nickel metal.

Long-period comet One taking thousands of years to travel around the Sun. On approaching the Sun it is observed as a 'new' comet.

Maria (lunar) Dry 'seas' on the Moon; dark basins covered by lavas.

Mesosiderite Stony-iron meteorite.

Meteor Light from a tiny particle burning up in the atmosphere.

Meteorite Natural object that survives its fall to Earth from space.

Meteoroid Potential meteorite; natural object in space that may land on Earth.

Micrometeorite Tiny meteorite; most are smaller than 1 millimetre.

Mineral Natural substance in which the atoms are structurally arranged and whose composition is fixed, or varies within limits. Minerals are the building-blocks of rocks.

Nebula An object in the universe that appears larger and fuzzier than a star. Most are glowing gas and dust; some are dark and seen in silhouette against a bright background.

Noble metal Metal, such as gold, resistant to acid attack.

Octahedrite Iron meteorite composed of intergrown high-nickel and low-nickel metal. See also 'Widmanstätten pattern'.

Olivine Stony mineral of magnesium and iron, common on Earth in basalt.

Oort cloud 'Cloud' of comets inferred to exist at the far reaches of the Solar System.

Organic compound Carbon bonded with hydrogen, nitrogen or oxygen once thought to be produced only by life.

Pallasite Stony-iron meteorite composed of olivine and metal.

Pyroxene Group of stony minerals of magnesium, calcium and iron, common on Earth.

Red giant 'Cool' star larger than the Sun.

Retrograde orbit Path of object round the Sun in the opposite sense to the Earth.

Shooting star Meteor; a streak of light in the sky.

Short-period comet Comet taking less than 200 years to orbit the Sun.

Solar wind Electrically charged particles from the Sun.

SNC meteorites Achondrites thought to come from Mars.

Stony-iron Meteorite with metal and stony minerals in roughly equal amounts by weight.

Stony meteorite One composed of silicate (stony) minerals, but may have up to 25 per cent metal by weight.

Supernova Exploding star that produces and distributes heavy elements.

Tektite Object of natural glass formed by impact on the Earth.

Terrae – see 'highlands'.

Terrestrial age Time since a meteorite landed on Earth.

Widmanstätten pattern Intergrowth of nickel-rich and nickel-poor metal, revealed after polishing and etching the surface of most irons.

INDEX